_____ your neurons are firing. *End of story*. But is this really so? Sharon Dirckx argues brilliantly that this is not the end of the story. The author combines her professional expertise with the clarity of a teacher to explain that we are more than machines. She claims, furthermore, that the question "Am I just my brain?" is not simply for the neuroscientist and philosopher. It has implications that affect all people. She gives compelling reasons why we should take the Christian message seriously. This work provides excellent food for the mind as much as for the heart."

Dr Pablo Martinez
Psychiatrist and Author

Laying out the arguments in her usual very readable style, Sharon makes a compelling case for why the answer to her book's title [spoiler alert!] is "No". Whether you agree with her conclusions or not, this whistle-stop tour of the hottest issues in neuroscience is a helpful, clear and concise summary of the different philosophical and theological positions, and the latest scientific data.

Dr Ruth M. Bancewicz
The Faraday Institute for Science and Religion, Cambridge, UK

In this fresh, clear, and helpful book, Dr Dirckx opens up a key part of what has been called "the most important conversation of our time". Is freedom only a fiction? Is human dignity merely a form of "speciesism"? Are we no more than our brains? The answers to such questions affect us all, and it is vital that we all explore them."

Dr Os Guinness
Author and Social Commentator

Books on this subject are often written by experts in philosophy and can be very difficult for the average reader to understand. This volume is written by a neuroscientist and is intended for non-specialists. The glossary and summary diagrams should make this important subject accessible to a greater number of people. I found the presentation to be both enjoyable and thought-provoking, and warmly commend it to you.

Dr John V. Priestley, Emeritus Professor of Neuroscience,
Queen Mary University of London, UK

Are we nothing more than the atoms of which we are made? Can humans be reduced to just the lump of grey matter between our ears? Sharon Dirckx draws upon her doctoral work in the sciences together with her years of experience in explaining the Christian faith, to help the reader think their way through this crucial question. Whether you're a Christian who wants to respond intelligently to new questions from neuroscience, or someone who suspects that the secular story isn't the whole story, *Am I Just My Brain?* will help you get not just your head—but also your heart, mind and everything else that makes you *you*—around this fascinating topic.

Dr Andy Bannister Speaker, Author;
Director, The Solas Centre for Public Christianity

Sharon Dirckx has written an excellent primer on the challenging subject of human consciousness. In this marvellous little book she has defined and discussed the major topic points with clarity, and skilfully makes difficult concepts easier to understand. The result is a solidly presented case for our minds being more than just our physical brains. It examines questions that neuroscience can't answer, such as *why* we can think, and shows how this ultimately points us to the reality of a creator God. Thoroughly recommended!

Dr Gordon Dandie FRACS
Neurosurgeon, Sydney, Australia

Dr Dirckx is well qualified to investigate the question that forms the title of her book. She illuminates the widespread reductionist notion that the brain and the mind are the same, and shows that it depends more on a presupposed naturalist or materialist philosophy than it does on actual science. This book is for the open-minded, and will enrich the reader whatever their worldview. I wholeheartedly recommend it.

John C. Lennox
Emeritus Professor of Mathematics, University of Oxford

This book shows how the perceived gap between God and the brain doesn't have to be a block—and can be a signpost. Sit at the learned feet of an experienced Christian neuroscientist and discover how...

Steve Adams
Author, *The Centre Brain*

Am I just my brain?

SHARON DIRCKX

Am I just my brain?
© Sharon Dirckx, 2019

Published by:
The Good Book Company in partnership with:
The Oxford Centre for Christian Apologetics and
The Zacharias Institute

thegoodbook.com | www.thegoodbook.co.uk
thegoodbook.com.au | thegoodbook.co.nz | thegoodbook.co.in

Sharon Dirckx has asserted her right under the Copyright, Designs and Patents
Act 1988 to be identified as author of this work.

A CIP catalogue record for this book is available from the British Library.

ISBN: 9781784982751 | Printed in Denmark

Design by André Parker

Contents

For my parents Dennis and Pauline.
Your love and support over a lifetime have
ultimately made this book possible.

Introduction

An early childhood memory of mine is of sitting by a window on a rainy day, watching the drops splash against the pane. Like all normal children, I spent most of my life racing around. But at this particular moment, I was still, and my mind had time to drift. I remember a series of questions popping into my head:

Why can I think?
Why do I exist?
Why am I a living, breathing, conscious person who experiences life?

I don't really remember where the questions came from. Neither do I remember my exact age. They were just there. Unprompted.

I know I am not the first to have this kind of "moment". When we sit still for long enough, all kinds of things bubble to the surface of our consciousness. Mindfulness gurus even tell us that bringing this kind of awareness into the foreground is good for our health. The more we are in touch with our inner life (such as our heartbeat, breathing and underlying emotions) and our outer environment (such as birds singing in the distance and doors slamming in the next room), the better. Conscious awareness seems to be central to what it means to be a living, breathing human being.

But what exactly *are* human beings? And how do we marry "aha" moments, such as the one described above, with some of the narratives coming from the sciences? Are we merely advanced primates? Are we machines? Are we souls confined to a body? Or are we some combination of all three? There are lots of different responses out there. Some of the loudest voices to answer this question come from neuroscience. They respond, "*You are your brain.* You are your neurons. *Why can you think?* Because your neurons are firing. End of story."

Francis Crick, who co-discovered DNA and won the joint Nobel Prize in Physiology or Medicine in 1962, said this in his book *The Astonishing Hypothesis:*

> *"You", your joys and your sorrows, your memories and your ambitions, your sense of personal identity and free will, are in fact no more than the behavior of a vast assembly of nerve cells and their associated molecules. As Lewis Carroll might have phrased it: "You're nothing but a pack of neurons". This hypothesis is so alien to the ideas of most people alive today that it can truly be called astonishing.*

Fifty years later, this hypothesis seems far from alien. In fact, many no longer consider it a hypothesis. According to them, it is the truth. The *only* truth.

Is Crick correct? Do our brains entirely account for who we are? How we answer this question has very far-reaching implications.

There are implications for free will. If we are driven by our brains, then are we really free to make decisions or are we simply driven by the chemical reactions within? On these grounds, how can anyone be held responsible for their actions, good or bad?

There are implications for robotics. Robots occupy more and more of the work force and have now entered our homes in the form of Google Assistant, Alexa and Siri. Will we eventually be able to manufacture conscious robots who are fully but artificially intelligent?

There are implications for ethics. If our brains define us, then personhood is dependent on having a fully functioning brain. But if that is true, then what status should we assign to those whose brains are not yet fully developed, such as premature and newborn babies? Or those whose brains have never functioned to full capacity, such as those with learning disabilities? Or those whose brains once functioned well but are now in a state of degeneration due to Alzheimer's disease or vascular dementia? In fact, none of us are exempt here. Beyond the age of 18, even a fit and healthy person begins to lose brain cells at an alarming rate. Our brains decline with age. Does this mean that personhood does too?

Finally, there are implications for religion. Since it has come to light that the brain is highly involved in religious belief and experience, can neuroscience now explain religion away? Is religious belief merely a brain-state, confined to those with the correct anatomy?

"Am I just my brain?" is not simply a scientific question. It taps into questions of identity that science alone cannot answer, and to fully consider the question we will need perspectives from philosophy and theology as well as neuroscience.

The mind is of particular importance in this conversation. Is there more to us than neurons because there is such a thing as the mind? We don't merely secrete brain chemicals; we also think thoughts. And we don't think with our brains but with our minds. But what exactly *is* the mind, and how does it relate to the brain? Herein lies the rub. The relationship of mind to brain is disputed. Essayist Marilynne Robinson, in her book, *Absence of Mind*, reads the situation well by pointing out that...

> *Whoever controls the definition of the mind controls the definition of humankind itself."* [1]

The answer you give to the question "Am I just my brain?" is not simply for the neuroscientist and philosopher. It has implications that affect all people.

Glossary

It's unavoidable that a book on this subject will include many specialist terms. I have tried to keep the technical biological language to a minimum, but the words that philosophers use to describe the ideas discussed can be just as confusing. Hopefully the following list will help you navigate the thoughts, questions and answers in this book a little more easily.

Compatibilism: The view that determinism is true but is also compatible with free will. Compatibilists believe that humans are determined by prior causes but can also act freely when they are not being constrained or are seeking to fulfil their desires. This is also known as "soft" determinism.

Consciousness: A property of the mind through which our subjective thoughts, feelings, experiences and desires have their existence.

Determinism: The belief that prior causes guarantee a particular outcome. Every event has a cause.

Downward Causation: The process by which the mind is able to act "downwards" on to the brain and cause changes in the brain.

HADD: Hypersensitive Agency Detection Device. A device that cognitive scientists of religion say is built into the human mind, enabling patterns, signals and other agents from the surroundings to be picked up.

Hard-Determinism : The belief that prior causes entirely guarantee a particular outcome, such that it could never have been otherwise. In neuroscience this equates to the belief that the human brain and the choices arising from it are determined on every level by prior causes, thereby ruling out the possibility of free will.

Incompatibilism: The view that free will and determinism are incompatible with each other, which can be held by hard determinists and libertarians alike, but for different reasons. Hard determinists believe that the fixed nature of the brain rules out the possibility of free will. Libertarians believe that the human will is free from constraints, and therefore the brain cannot be determined on every level.

Libertarianism: The view that freely made decisions can be made by agents (here, people) that are not determined by prior causes. This view upholds human free will.

Materialism: The view that observable matter in time and space is all that exists. For the purposes of this book, it is used interchangeably with physicalism.

Mind: The bearer of the unseen, inner life of a person, in the form of thoughts, feelings, emotions and memories. The mind is the bearer of consciousness.

Neurologist: A physician who is trained in diagnosing and treating disorders of the brain and nervous system.

Neuroscientist: A scientist who studies the brain and its functions.

Neurosurgeon: A physician who is trained in diagnosing and performing surgery on patients with disorders of the brain and nervous system.

Non-Reductive Physicalism (Neuroscientific): The view that the mind has been generated by the brain. When a number of component parts come together and reach a certain level of complexity, something new (the mind) emerges. The mind is physical but cannot be reduced to physical processes alone.

Physicalism: The view that the observable physical world is all that exists. For the purposes of this book, it is used interchangeably with materialism.

Psychiatrist: A physician who is trained in diagnosing and treating those with mental illness. A psychiatrist is able to prescribe medication as part of a patient's treatment.

Psychologist: A non-physician who is trained in treating those with mental illness. A psychologist is not able to prescribe medication and is likely to treat patients by training them in mental exercises.

Reductive Physicalism (Neuroscientific): The view that the mind is reducible to physical processes in the brain. Therefore, there is really no such thing as the mind. The mind is the brain.

Substance Dualism (Neuroscientific): The view that two distinct substances characterise the mind-brain relationship: a physical brain and a non-physical mind. The mind can exist without the brain but in humans they interact. The mind is beyond the brain.

Am I really just my brain?

I will never forget the day I saw a human brain removed from a corpse. At that moment, I was already very familiar with the human brain, having spent years imaging and studying it. Yet, this experience was different altogether.

A group of us, dressed in green robes, wearing blue plastic shoes, were in a dissection room in a medical school. The icy formality matched the cold air of the surroundings. The pungent smell of formaldehyde, used to preserve human tissue, filled our nostrils. The body of an older woman lay on the bench before us.

This was not the first time I had seen a corpse, but there was something different about this setting. The woman had donated her body to medical research. We were there to study the anatomy of the human brain, and the first stage was to watch its removal from the body. Our anatomy professor and instructor began. There was no blood involved as the person had died some time ago, but a lot of sawing and, at times, brute force to cut around the skull to expose the brain. Despite the clunky technique, it was a deeply sobering and reverent experience, conveying utmost respect for the unnamed woman who had given her body so that others could learn.

A few minutes later, and there it was in its entirety. A mass of water and fat, weighing just 1.5kg (3.3 pounds). I went into study-mode thinking less about the person and more about the anatomy of the brain. Yet, it was undeniable that on the table in front of us was the mediator of the thoughts, feelings, longings and experiences of this unnamed woman.

***

To the touch, the human brain is not unlike the consistency of mushroom. Yet, mercifully, you do not have mushroom between your ears. Quite the opposite. This incredible organ comprises just 2% of the body's weight, yet it uses 20% of its energy, despite being nearly 75% water. The human brain contains roughly 86 billion brain cells known as neurons. Each of those neurons can send up to 1000 nerve impulses per second to tens of thousands of other cells, at speeds of up to 430 km/h (268mph).[2] As you are reading these words, your brain is generating enough electricity to power an LED light, and every minute enough blood flows through your head to fill a wine bottle. The human brain is more developed than in any other creature, although the prize for the biggest brain goes to the sperm whale, weighing in at 7.5kg (17 pounds).

Every thought, memory, emotion and decision you make is filtered through this thing known as your brain. Changes to the chemistry and physiology of our brains affect our capacity to think. For example, just a small amount of dehydration can dramatically affect our attention span, our memory, and our ability to think clearly. And many of us know that a morning shot of caffeine is vital to kick-start our thinking processes at the beginning of a new day.

But we now also know that changes to our thinking impact the brain itself. Scientists used to think the brain was fixed and rigid, but it is now known to be incredibly "plastic", in the sense that it is constantly changing and forming new connections and pathways throughout a person's lifetime. Changes to the brain affect our thinking. But our thinking, our lifestyle and our habits also shape the way our brains grow and develop.

STUDYING THE BRAIN

From a young age I knew I wanted to be a scientist. I worked hard at school—probably a bit too hard—and in my early teens I was already dreaming of doing a PhD. School in Durham led to university in Bristol in the UK, where I studied biochemistry..

I loved the lectures, but was less keen on the lab work. In my day, biochemistry labs were warm places, often with a strong yeasty smell. Students in white coats could be found blending, spinning or shaking exotic concoctions, pipetting tiny amounts of liquid from one test tube to another, or watching anxiously while their glass flasks enjoyed a long, hot soak in a bath. It could be weeks or sometimes months before we discovered whether an experiment had worked. And if it had not, it was time to start again. This was in the mid-1990s. Things have moved on since then.

It was in Bristol that I first heard about brain-imaging. Some friends studying physics were trying to squeeze results out of an archaic machine more or less held together with parcel tape, just down the corridor from my research lab.

They were using what was then a new technology that enabled them to look inside the body without making a single cut: magnetic resonance imaging (MRI). I was drawn to this technique and began a doctorate at Cambridge University two years later. I remember clearly the four-year-old daughter of one of the researchers reminding us of the unique selling point of MRI: *"Daddy, doesn't it hurt when they slice into the man's brain like that?"* She was watching a screen showing a rotating man's head and slices gradually peeling off to show more and more of the inside of the brain. Does it hurt? Not one bit. With MRI you get electronic slices of brain, not real ones.

One of the most exciting contributions of brain-imaging is that it enables scientists to study the brains of healthy people. At the turn of the 20th century, when the only way to see inside a brain was to pick up a knife and start cutting, the only subjects available for investigation were those with sufficiently unpleasant or incurable diseases that they were willing to try anything; or else those in whom the disease had already run its full course. The arrival of brain-imag-

MAGNETIC RESONANCE IMAGING AND ITS ABILITY TO SEE
INSIDE THE HUMAN BRAIN

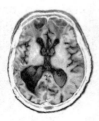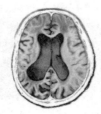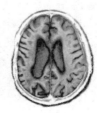

ing techniques meant that healthy and diseased brains could now be compared.

Fast forward to the 1990s, and functional MRI (fMRI) took imaging to another level by enabling us to look not just at the structure in a series of static images but also at brain *activity*. Recall the times when you have climbed a tower, where the effort of ascending is rewarded with a spectacular view. At the top, our gaze often focuses first on the larger, fixed and easily recognizable structures such as buildings and streets. But then we also notice movement from pedestrians, cars and buses. MRI is most commonly used today to look at fixed anatomy in the brain or other parts of the body, such as knee or shoulder joints. By contrast, fMRI measures *movement* inside the brain, specifically the movement of blood. When part of the brain starts working harder, more blood rushes into it bringing supplies of oxygen and sugar. Functional MRI measures that blood-flow and can tell us which part of the brain is at work. Its development in the late 1980s shaped the landscape of neuroscience for decades to come—a landscape we are still exploring today.

I had the privilege of spending eleven years in fMRI research and worked with some brilliant neuroscientists who have made significant contributions to this field of research. Through fMRI, we investigated how the brain can reorganise itself around a tumour, or become taken over by an addictive drug. At the outset, my research focused on healthy volunteers, but I later went on to work with cancer patients and cocaine addicts as well.

ARE WE JUST OUR BRAINS?

As I think about the body in the dissection room, which used to be a living breathing woman, I can't help but ask, "What is it that makes me a person?" Many answers are offered today. The fashion and cosmetics industry says, "You are your body". The financial world might say, "You are your income". Politicians say, "You are your influence". The academy would say, "You are what you write". More recently, neuroscientists have started saying, "You are your brain". To understand a person is to understand their brain. To understand the brain is to understand the person.

What are we to make of this view? According to the "you are your brain" view, neuroscience can now speak to the fundamental question of human identity. For some, neuroscience has become the lens through which we make sense of all areas of life. Brain maps have been used to make marketing decisions, economic decisions, even legal decisions. Rather than asking someone's opinion, we scan their brain! Professor Raymond Tallis, a retired clinician and neuroscientist at Manchester University, has described this as "neuromania".[3] Neuroscience has made astonishing discoveries that have advanced our understanding and our ability to diagnose and cure diseases. But have we also become obsessed with the possibility that it could answer every question we have?

WHERE DID THIS START?

Before we tackle the heart of the question, it can help to understand where this viewpoint has come from. At first impression, the belief that the brain accounts for everything appears new, as though it has been forced into being through the rise of neuroscience. However, this belief can be traced back to ancient Greece, and in particular to the

5th century BC. The physician Hippocrates (460-377 BC) is best known for his Hippocratic oath which can be summarised as *do no harm*, but he also studied and wrote on epilepsy. In his work *On the Sacred Disease* he commented (my emphasis added):

> *Men ought to know that **from the brain, and from the brain only**, arise our pleasures, joys, laughter and jests, as well as our sorrows, pains, griefs and tears.*[4]

Hippocrates was making the point that epilepsy is not caused by demon possession, as was commonly thought at the time, but that it is a disease of the brain. Yet this phrase, *from the brain and the brain only*, has shaped a growing modern viewpoint that "mind equals brain" in every way.

This view has been expressed in academia in recent times through people such as Sir Colin Blakemore, Professor of Neuroscience at the University of Oxford, who in 1976 said:

> *The human brain is a machine which alone accounts for all our actions, our most private thoughts, our beliefs. All our actions are the products of the activity of our brains.*[5]

And views espoused in the academy eventually filter into popular culture. The Disney animation *Inside Out* is one such example. The movie creatively depicts the complexity of the human brain *and* the importance of different emotions (Joy, Sadness, Anger, Fear and Disgust are all characters in the movie). Plasticity in the brain is depicted as the breaking and reforming of various "islands". However, the narrative thread running through this movie is that everything that makes the protagonist, Riley, who she is, comes from physical mechanisms inside her head. When Riley's core memories and "islands" are intact, her outward behaviour is balanced. When they are not, the outside world falls apart. As the title suggests, there is only inside out. There is no outside in.

WHERE SHOULD WE START?

How should we begin to examine this question, "Am I just my brain?" A helpful starting point is to be open to the possibility that it cannot be answered by neuroscience alone. At first glance, this question appears to be scientific in nature, primarily because the question is raised by scientists and references a part of our anatomy. But in fact, "Am I just my brain?" asks a philosophical question about human identity. Neuroscience alone is unable to answer these kinds of questions. Neuroscience describes what is going on in the brain in beautiful detail, and is the obvious go-to discipline to answer questions like "What is a brain?" and "How does the brain work?" But the question "What is a person?" is very different. It reaches beyond the scientific method into philosophy, ethics and, many would argue, theology.

Human memory has many different components, one of which is your working memory—essentially the "notepad" in your head. Working memory is the part of your brain you use when trying to remember the shopping list you scribbled down earlier and then left at home. Imagine if a neuroscientist studying human working memory decided that they were only going to refer to results from functional MRI and ignore all other disciplines such as physiology, anatomy and pharmacology. Frankly, this would be poor scientific practice, leading to a diminished understanding of working memory. A good scientist uses all of the tools at their disposal and seeks to set their results within the wider context of other disciplines. Similarly, to try to answer questions of human identity using neuroscience alone is to sell ourselves short. We need to reach beyond neuroscience to answer questions of identity. Neurons and brain chemicals alone will not get us there and will leave us with a diminished view of the human person. "Am I just my brain?" is not simply a scientific question but also a philosophical one, so our journey will traverse philosophical terrain as well as that of brain science.

At the heart of being a good scientist is the need to be open to new ideas and unexpected results. A common understanding of science involves things like setting a hypothesis, data collection and the interpretation of the data. The hypothesis is our theory of what we expect to observe. If the data fit the hypothesis, then we may be on to something. The next stage is to attempt to repeat the results. If we succeed several times, then the hypothesis begins to look as though it is correct. If, however, the data do not fit the hypothesis, we need to be open to the possibility that our hypothesis is wrong and needs revising.

Among scientists there is sometimes a temptation to "fudge" the data to make it fit the hypothesis. Yet, some forward leaps in science have come through unexpected results and the courageous revision of long-standing theories in the face of criticism. The need for an open mind is crucial to the success of a scientist. I want to invite you to apply the same open mind to the topics we are discussing in these pages.

ARE YOU OUT OF YOUR MIND?

So far, we have been discussing the brain—the mushroom-like structure between your ears, consisting of millions of interconnected neurons awash with chemicals, hormones and electrical activity. But we don't simply possess neurons, we also think thoughts. We seem to also have a mind. So what exactly is the mind?

The *Merriam Webster Medical Dictionary* defines "mind" as:

> *The element or complex of elements in an individual that feels, perceives, thinks, wills, and especially reasons.*

NEURONS ARE THE BUILDING BLOCKS OF THE BRAIN
They join together at synapses.

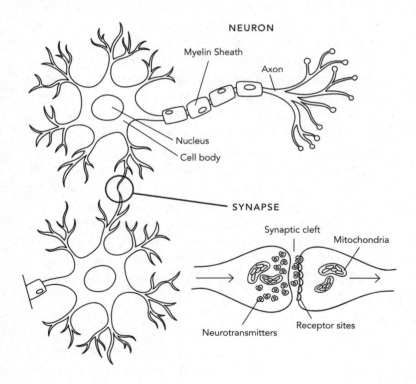

NEURON

Myelin Sheath

Axon

Nucleus

Cell body

SYNAPSE

Synaptic cleft

Mitochondria

Neurotransmitters

Receptor sites

The Oxford English Dictionary defines it as follows:

The seat of awareness, thought, volition, feeling, and memory.[6]

In other words, the mind is the bearer of the unseen, inner life of a person, in the form of thoughts, feelings, emotions and memories. When you select a playlist from your phone, recall a conversation from the day before, or experience a hurtful comment on social media, your mind is engaged.

What then is the connection between the brain and the mind, between neurons and thoughts, between synapses and sensations? How do you get from brain voltages to "I'd like to play tennis today?"

Mind and brain are clearly related. But how exactly? This, is the million-dollar question that lies at the heart of this book. This conundrum has occupied philosophers, ethicists and theologians for centuries. Many different answers have been offered to what is known as "the mind-brain problem".

THE MIND-BRAIN RELATIONSHIP
How do we get from neurons to thoughts?

WHAT ARE THE OPTIONS?

One popular modern view, the view under scrutiny in this book, is that the mind *is* the brain. Mind and brain are identical. Thoughts, memories and emotions *are* the firing of neurons. No more. No less. This view is sometimes referred to as "reductive physicalism". The mind is reducible (hence "reductive") to the physical workings of the brain (hence "physicalism"). In other words, there isn't really such a thing as the mind, but only the activity of the brain.

The voices that espouse this view are loud but are by no means the only ones in the choir. There are several alternative descriptions of the mind-brain relationship in circulation today that thinking people believe to be viable and persuasive. These views espouse a distinct mind that may interact with the brain but is certainly not at the mercy of it. In this book, I want to demonstrate that "you are your brain" is far from the only option available.

One alternative view is that the brain *generates* the mind. When the components of the brain combine and reach a certain level of complexity, they give rise to something new and distinct: the mind. This view is referred to as non-reductive physicalism (NRP). The mind arises from the *physical* brain (hence "physicalism"). But once formed, this new entity cannot be reduced back to its original components (hence "non-reductive"). But if those components are broken up, the new entity goes away.

We might summarise this view as "The whole is greater than the sum of its parts".* 7 According to this view, the

---

* Some philosophers define NRP differently, such as William Jaworski *Philosophy of Mind: A Comprehensive Introduction* (Wiley-Blackwell, 2011). Different scientific domains cannot be reduced to each other; for example, biology is not reducible to chemistry. However, according to this view, the building blocks of consciousness will always be non-conscious matter.

THE MIND-BRAIN PROBLEM: THREE OPTIONS

THE MIND *IS* THE BRAIN
Reductive Physicalism

THE BRAIN *GENERATES* THE MIND
Non-Reductive Physicalism

THE MIND IS *BEYOND* THE BRAIN
Substance Dualism

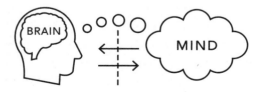

mind is more than the brain but is inextricably bound to it. One obvious question raised by this view is this: *when the brain dies, what happens to the mind?*

A second alternative is that the mind *is beyond* the brain. Mind and brain are two distinct substances that interact but can also operate independently of each other. This view is known as "substance dualism" because two substances are involved in the mind-brain relationship: a physical brain and a non-physical mind.

A question raised by this view is this: *how exactly does a non-physical mind interact with a physical brain?* Especially since neuroscience shows a strong connection between the two.

In chapters 3 and 4, we will consider and critique these mind-brain descriptions and others, through the lens of consciousness. However, the scientific method alone will not be enough to help us in our quest; we need to look at the different beliefs that people bring to their science, and, in fact, to the whole of life. All people have beliefs, including scientists, but we need to understand the nature of those beliefs if we are to see how to synthesise a view of the world that has integrity.[8] One way to test a belief is to ask the following three questions.[9]

1. Is it internally coherent?
Does "you are just your brain" make sense according to its own frames of reference? Is it a watertight position, or are there internal inconsistencies? Aristotle[10] (c.384-322 BC) made the point that beliefs that only allow for physical things undermine the scientific method. The goal of a scientist is to make sense of the physical world. But if we are merely of the same composition as the world we study, then

how is it possible to make any form of objective claim? Is "you are just your brain" internally coherent? Not at all. It undermines the very discipline that its proponents practice and applaud: science itself. We will also see in the subsequent chapters that this view even calls the integrity of the human mind into question.

2. Does it have explanatory power?

Does "You are just your brain" explain the world around us? Does it make sense of the world we live in? If something is true, then it ought to help us make sense of the world rather than throw us into further confusion. Is this true of the view that a person *is* their brain? When I think of what it is that makes me who I am, neurons alone seem insufficient.

A large part of who I am comes from an unseen inner life consisting of thoughts, memories, emotions and decisions, none of which are captured by cell voltages, neurotransmitters and blood-flow changes. "You are just your brain" instinctively fails to explain the inner "me".

3. Can it be lived?

Francis Schaeffer (1912-1984) founded l'Abri in 1955, a Christian community in the Swiss Alps, and a haven for those asking searching questions. One of Schaeffer's convictions was that the extent to which a belief could be authentically lived out and lined up with our experience of life is a marker of its truthfulness.

And what is our experience? We live as though *we* do the thinking, not our brains. Neurons do not think: *people* think. We live all the time as if there is such a thing as a first-person perspective of the world.

Mindfulness, self-help, counselling, autobiographies,

child-abuse scandals, or indeed anything that requires introspection, all assume that the first-person vantage-point is real. We live as if there is far more to us than simply our brain.

If the answer to "Am I just my brain?" is, "no", then what more is there to me? In the past, we commonly referred to the part that is essentially "me" as the soul. Is there such a thing as the soul and, if so, does it help us answer fundamental questions of human identity? Of course, some believe that the soul can now also be explained away by neuroscience; in other words, belief in the soul is out of date. Is this true? That's the question we turn to now...

Is belief in the soul out of date?

American physician Dr Duncan MacDougall (1886-1920) is known as "the man who tried to weigh the human soul".[11] Tuberculosis patients whose lives were hanging in the balance were themselves balanced upon scales and weighed at the moment of death. As they heaved their last breath, MacDougall was poised and ready. He later wrote:

> The instant life ceased the opposite scale pan fell with a suddenness that was astonishing—as if something had been suddenly lifted from the body. Immediately all the usual deductions were made for physical loss of weight, and it was discovered that there was still a full ounce of weight unaccounted for.[12]

MacDougall concluded that his patients were 21 grams or one ounce of weight lighter after death, and that this must be the weight of the soul. The results were later discounted,[13] even though the concept lives on through Dan Brown's novel *The Lost Symbol* (2009) and the film *21 Grams* (2003).[14]

The soul features regularly in everyday conversation. We eat soul food, we have soul mates, we describe restaurants and concerts without an atmosphere as soulless. At pivotal moments in life, we take time for soul searching. In music, Aretha Franklin will forever be the Queen of Soul. On one level, reference to the soul is part of daily life. The soul has also had historic significance. For centuries, belief in the existence of the soul—the essence of a person—has had a pivotal role in the development of ethics and discussions on human dignity and rights. Many laws that we now think are self-evident, such as those promoting equality and prohibiting slavery and the mistreatment of prisoners, grew out of a conviction that all people have a soul.

Yet some scientists and philosophers believe the soul is now an out-of-date concept. In the words of philosopher Daniel Dennett, the soul has "*outlived its credibility thanks to the advance of the natural sciences.*"[15] Historically, according to this view, the soul filled a gap in understanding that has now been plugged by the sciences. The soul, if it exists, belongs to the physical realm. As reflected in Dr MacDougall's experiments, it has mass and can be weighed. Harvard psychologist Steven Pinker puts it more explicitly: he suggests that the soul is, in fact, the brain:

> *The supposedly immaterial soul, we now know, can be bisected with a knife, altered by chemicals, started or stopped by electricity, and extinguished by a sharp blow or by insufficient oxygen.*[16]

So, we see that the same arguments against the mind have also been levelled against the soul: that neuroscience not only explains mind and soul but also explains them *away*. The "You are your brain" view resurfaces. What should we make of this view? Before we can give an answer, we need to ask, what exactly *is* the soul?

THE ANCIENT SOUL

People have believed in the existence of the soul for thousands of years, and the story of the soul is often thought to begin in ancient Greece. The philosopher Plato (c.428-347 BC), who was heavily influenced by his mentor Socrates (c.470-399 BC), is well known for his views. The Greek word for soul is *psyche*, derived from the verb *psychein*, which means "to breathe" but was understood to be "the essential life of a being".[17] According to Plato, the soul is the ultimate source of life for all living things. For humans, the soul is *the essential person*. Plato argued that the soul occupies and gives life to a body, but it does not need the body and can exist without it. When the body dies, the soul lives on. At its best, the body is a carrier or container for the soul, but at worst, it serves as a prison, exerting evil effects upon the soul.

Other Greek philosophers, such as Pythagoras (c.570-495 BC)—who gave his name to the mathematical theorem we all had to learn in school—took a different view on the soul. Pythagoras believed in reincarnation: the belief that the soul is not necessarily tied to one particular body and can go on to occupy different bodies. Belief in reincarnation is central to Eastern thought and is held by billions of Hindus, Buddhists and Sikhs around the world today.

The 2016 award-winning animation *Kubo and the Two Strings* illustrates the central idea. When Kubo's mother suddenly dies, her 12-year-old son joins forces with Monkey and Beetle on a journey to solve the mystery of his family. He later discovers (spoiler alert) that the soul of his mother has been transferred to Monkey. The talking primate accompanying him has in fact been his mother all along. A soul can occupy different bodies at different times.

Aristotle studied under Plato and agreed that the soul exists and gives ultimate life to the body, but he came to different conclusions on how it interacts with the body. For Aristotle, the soul and body are integrated and together make the person. A particular soul occupies a particular body. The body is more than simply a carrier for the soul, and both body and soul need each other for their own existence. For Aristotle, a person is an amalgamation of body and soul, or "matter" and "form". Just as the four-legged object in your kitchen is made of wood and also has the form of a table, so a person is made of matter and has the form of a soul. This position is known as hylomorphism, from the Greek, *hylē*, meaning "matter", and *morphē*, meaning "form".

THE MODERN SOUL

We encounter ideas about the soul in many different ways today. Google "mind, body, soul", and you will be plunged into a world of spa resorts, yoga classes, protein shakes, detox and all things associated with general wellbeing. Perhaps we can agree that a leisurely day of seaweed wraps and thermal baths does something beyond simply exfoliating and toning our bodies. The subtext of all these products, experiences and messages is that we achieve wellbeing at the deepest level by attending to this entity called the soul.

The Christian faith also has a lot to say about the soul. The soul is that which makes us more than matter, more than advanced primates, more than simply our brains. The soul is the impenetrable core of a person, given by God. We can either feed our soul or starve it. In the Bible, the word "soul" is used in different ways. At times it designates the whole person: for example, when the psalmist says, "Praise the Lord, O my soul" (PSALM 104 v 1). At other times, "soul" refers to one part of a person, distinct from body and spirit, such as in the closing words of a letter from the apostle Paul:

> *May your whole spirit, soul and body be kept blameless*
> *at the coming of our Lord Jesus Christ.*
> 1 THESSALONIANS CHAPTER 5 VERSE 23

Jesus teaches his disciples about the soul as they ponder the danger they will later face for being his followers. He tells them that even though their bodies may be killed, their souls are untouchable in this life, but are dependent upon God for their continued existence in the life to come. Jesus says:

> *"Don't be afraid of those who want to kill your body;*
> *they cannot touch your soul. Fear only God, who can*
> *destroy both soul and body in hell."*
> MATTHEW 10 v 28

The soul is a complex phenomenon in the Christian Scriptures and cannot be boxed into a single idea. It comes as no surprise, then, that there are different views on the soul within the Christian tradition, which we will say more about later. But whatever each Christian's precise view is, all are agreed that humans are more than simply physical beings.

The language of Plato's "immaterial and eternal soul" was imported into Christianity through the 4th-century theologian Augustine of Hippo (AD 354-430), and humans were subsequently thought of as immaterial souls floating temporarily within material bodies from which they will eventually depart to be with God. This belief was the predominant Christian view for the following 1,000 years, and still holds great sway today, but it owes more to Greek philosophy than it does to Jesus' teaching or the rest of the Bible.

The French philosopher René Descartes (1596-1650) introduced Augustine's ideas into modern philosophy and shifted the conversation from "soul" to "mind". Descartes wrestled with the question of what he could be certain of in life and arrived at this conclusion: he could be certain of his own mind. Above all else, he could be certain of his own inner life of thoughts, doubts and questions. "I doubt; therefore I exist" became the iconic expression...

Cogito ergo sum: I think; therefore I am.

For Descartes, the mind was primary and the body secondary. So, how do the two interact? He concluded that they interact through the pineal gland, a structure buried deep within the brain. Descartes' divorcing of body from mind became known as Cartesian dualism, often parodied with the phrase "the ghost in the machine".

NOTHING NEW UNDER THE SUN

It might appear that belief in an entirely physical soul, reducible to physics and chemistry (the view known as reductive physicalism) is new and unique to the neuroscientific age in which we find ourselves. However, this belief has existed since ancient times. Plato, Socrates and Aristotle defended their views against physicalist views held by their contemporaries.

The philosopher Democritus (c.460-370 BC) was the

first to put forward the concept of a universe composed of atoms. He described the soul as "a kind of fire made up of atoms", a view that was later supported by Epicurus (341-270 BC) and Lucretius (1st century BC). Plato and Aristotle were accustomed to discussing ideas on the soul and responding to the claims of Democritus. So the idea of a material soul is far from new.

ARE WE GHOSTS OR MACHINES?

Many of the debates surrounding God and science give the impression we must choose between two mutually exclusive, polar-opposite positions. Regarding the soul, it can seem as though we must opt for either disembodied ghosts or robotic machines, with no alternatives in between. The soul is either a nebulous immaterial entity that has little relation to the body or else the soul has been displaced by neuroscience and psychology and is, in fact, the brain. Yet, neither description seems to capture both the conclusions of great minds who have thought deeply about this issue over millennia and our experience of ourselves as living, breathing, thinking people. Both options sell us short. One purpose of this book is to show that the choice between "ghosts" or "machines" is a faulty one. There are more options available: views that occupy a middle ground that make more sense of people, science and daily life. Views that are held by both theists and non-theists alike.

Some theologians and philosophers today still consider the soul to be the immaterial core of a person that integrates everything else: the mind, the will and the body. Some lean towards Augustine's view, but others are drawn to the more holistic views of medieval theologian Thomas Aquinas (1225-1274).[18] Aquinas was influenced by Aristotle and took a more integrated approach to soul and body. According to Aquinas, the soul "moves" the body to act but also needs the body for its own flourishing. The soul can survive without the body but it is incomplete. Theologian Eleanor Stump likens it to "the shell of a house before it is fully built".[19] The shell can stand on its own but it is not yet fit for purpose. Aquinas' views are known as Thomistic dualism or hylomorphic dualism.

Other theologians believe that the soul should be described in material terms but as something that emerges to be greater than the sum of its parts.[20] According to this view, the soul is not a distinct part of the person, separate from the body, but a property that has emerged alongside our human capacity to relate to other people and to God. This view differs from those of Dennett and Pinker because the soul here is dependent upon, but not reducible to, the high levels of brain complexity seen in humans.

A brief glance at philosophy, then, shows us that "ghosts" and "machines" are far from the only ways to describe the soul. There are many alternatives that do a better job of explaining and integrating the two extremes.

YOU HEARD IT HERE FIRST

Many assume that Plato was the first to mention the soul. Yet discussion of the soul goes back even further than ancient Greece. Ancient Hebrew texts such as the Old Testament, dating as far back as 2,000 BC, use the word *nephesh* over 750 times. *Nephesh* has a variety of meanings, but in relation to people it refers to the essential core of the person, both physical and non-physical. In some instances, this means the vital entity that makes and keeps someone alive in a very literal sense, or that which underpins the emotions, will and desires as expressed in the mandate to,

> *...love the Lord your God with all your heart, and*
> *with all your soul [nephesh] and with all your strength.*
> DEUTERONOMY 6 v 5

This verse is also quoted in the New Testament, in which *nephesh* translates in Greek to "psyche". The English language is limited to the use of a single word, "soul"; however, a variety of different Greek words exist. If a strictly biological meaning had been intended, then a different word, "*bios*", would have been used in the Bible instead of *psyche*. The use of *psyche* suggests that the Bible is capturing something that perhaps includes the physical or biological but also goes beyond it.

MIXED MESSAGES

The view that there is no soul comes largely from voices within neuroscience and philosophy who believe that we are living in a solely material world.

But if we consider, for example, the conversation surrounding human sexuality, we see the soul being referred to very differently. Caitlyn Jenner, formerly American athlete Bruce Jenner, wrote autobiographically in the recent book, *The Secrets of My Life*,

> *The juxtaposition of Bruce and Caitlin is shocking even to me. How could one become the other and the other become the one? I know that Caitlin was my gender identity at birth, waiting for the right moment to subsume Bruce. Imagine denying your core and soul. Then add to it the most impossible expectations that people have for you because you are the personification of The American Male Athlete.[21]*

For Jenner, and for many others, the soul is the core of the person and is the place where true identity lies, regardless of what is happening to the body. The way to fulfilment is to bring the body in line with the soul.

My aim here is not to delve deeply into the conversation on gender identity, except to say that as a society, profoundly contradictory messages are coming from different corners. One says, "You are no more than your body and brain"; another says, "There is so much more to you than your body and brain". There is something deeply confusing here. Some secular scientists say that the core of a person is physical. Many transgender advocates say things which seem to imply that the core of a person is non-physical; there is a soul that cannot be denied. Surely both of these cannot be true—so, which is it?

To help us move forwards, it will be helpful to look at some of the different descriptions of the relationship of the mind to the body. We will do this in chapter 3 by looking through a different lens; we will consider the mind-brain problem in more detail by asking, "are we just machines?"

Are we just machines?

Some of Arnold Schwarzenegger's best-known films are *Terminator* and *Terminator 2*, depicting how conscious robots would eventually overcome and obliterate humanity. 21st April 2011 was prophesied as being judgment day—the day when the robots would finally rise up and bring about the end of the human race.

The date has come and gone, along with numerous other apocalyptic calendar suggestions, and yet the question of whether machines can possess human levels of intelligence is more relevant than ever. The technology may not have reached the levels that director James Cameron predicted, but machines are increasingly taking on roles that used to be occupied by humans and, in many instances, now go well beyond human capacity.

The idea of artificial intelligence (AI) emerged in the late 1940s. Alan Turing is often thought of as the father of modern AI, having developed his ideas while deciphering the Enigma code during World War II.

In 1997, the IBM computer Deep Blue beat the reigning world chess champion Gary Kasparov in a six-game match, receiving media coverage across the world. Today, robots even have learning capacity. In 2017, Google's AlphaGo Zero mastered the ancient Chinese board game Go with no human help other than being told the rules. Within three days, AlphaGo had begun to invent rules of its own, and shortly afterwards the South Korean Grandmaster, Lee Sedol, was toppled.[22] The score? 100 to 0.

Some have even deemed androids to be acceptable marriage partners. In 2017, a Chinese AI engineer gave up on his search for a human wife and married a robot he had built himself with plans to "upgrade" her in the coming months.[23] Questions of AI no longer belong simply to the realm of science-fiction films; they are part of modern life. But these advances in technology leave us with questions. What is human intelligence? Does it differ from android intelligence? If so, how?

TESTING FOR INTELLIGENCE

Anyone with an iPhone will probably have had at some point a conversation with Siri, the "person" that resides in our hand-held devices. My children have even made attempts to strike up a friendship with their "electronic assistant".

Siri, how are you?
I'm fine. Thanks for asking.

Siri, who are you?
I am Siri.

Siri, how are you feeling today?
I'm happy to be alive!

Siri, who made you?
I, Siri, was designed by Apple in California.

Siri, what are you?
I'm Siri. But I don't like talking about myself

Siri, are you a person?
Sorry, Sharon, I've been advised not to discuss my existential status.

Siri, why are we here?
I don't know. Maybe the genius bar folks can answer that.

Siri, why do we exist?
To have conversations like these.

Siri, why do you exist?
I process; therefore I am.

In 1950, Alan Turing devised a method of measuring the level of intelligence of a robot,[24] known as the Turing test. An android and a human are anonymously asked a series of questions by human interviewers. The task of the interviewer is to examine the answers and distinguish between man and machine. If the level of complexity in the answers given by the android is indistinguishable from those of a human, then they have passed the Turing test and are considered intelligent. Today, some robots are barely distinguishable from humans. An android made by Russian and Ukrainian engineers convinced a third of its Royal Society interviewers that it was a 13-year-old Ukrainian boy[25] called Eugene Goostman. What is the significance of this kind of result? Does it mean that human beings are merely skin-covered machines and therefore, ultimately, entirely interchangeable with artificial entities?

CHINESE ROOMS

In short, not necessarily. Philosopher John Searle proposed a thought experiment, known as the Chinese Room, to suggest that the Turing test alone cannot guarantee intelligence.[26] Imagine person A sitting in a room passing Chinese messages through a narrow slot to Person B.

JOHN SEARLE'S CHINESE ROOM THOUGHT EXPERIMENT

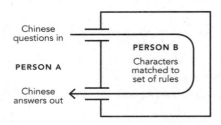

Chinese questions in

PERSON A

Chinese answers out

PERSON B

Characters matched to set of rules

Person B is fully enclosed and has no knowledge of the Chinese language, However, they have a table of Chinese symbols with which they can decode each message and construct a reply. To Person A, it seems as if Person B speaks Chinese, but this is incorrect. An ability to process Chinese characters is not the same as speaking the language. Similarly, an android's information-processing ability does not necessarily mean that they are conscious in a way that is meaningful and comparable to human consciousness. It is one thing for Siri, Alexa or "Eugene" to output words that make sense to us, but that are driven by complex mathematical algorithms. It's quite another thing for them to be self-conscious and understand what they are saying.

Perhaps the most pertinent question is this: do androids

have the *potential* to become conscious? As robots become more and more "intelligent", will they eventually take on a mind of their own and become indistinguishable from humans? The answer to this question centres around another question that has taken centre stage in recent years. What is the nature of human consciousness? Are we merely complex computers or is there more to it?

ARE YOU STILL CONSCIOUS?

It's Monday morning. I am in a local coffee shop, queueing for a flat white. There are several people in front of and behind me. Music is playing in the background, and I can just about hear it over the hum of the conversations. A child is having a tantrum in the corner. But it's okay, the smell of coffee reaches my nostrils. The Monday morning trip to the coffee shop is more than simply an act of acquisition; it is an experience of sights, sounds, smells and sensations all rolled into one. It is an instance of consciousness.

What exactly is human consciousness? Consciousness is hard to define. Consciousness is everything on your radar, both internal and external. Consciousness is the reason for the first-person perspective and the inner narrative in your head. Consciousness is why you integrate everything at the coffee shop into a single experience: the aroma, the screaming, the music. The unhindered off-loading onto people or paper at the end of a stressful day is a "stream of consciousness". Yet, through consciousness, you are aware of being the same person today as you were yesterday and will be tomorrow. Descarte's "I think; therefore I am" is an expression of consciousness.

As philosophy Professor Thomas Nagel puts it, to be conscious is for there to be something "that it is like to be us" that even someone who knows you well cannot speak to in quite the same way.[27] Each person has an inner world, an inner thought-life that is unique to them. How does consciousness relate to the mind, you might ask? Philosophers argue that consciousness is a property of the mind. The mind is the bearer of consciousness.

Philosophers use the term *quale* to describe a quality or property as it is perceived or experienced by a person. In simple terms, *qualia* (that's the plural) refer to "what something is like". Let's return to the coffee shop again. There is nothing quite like the smell of a rich Guatemalan blend. Even people who don't like coffee often admit to enjoying the aroma. But if someone asked you to *describe* the smell of coffee, how would you respond? It is an experience that cannot be reduced any further. If you want to know what coffee smells like, you need to smell it! Life is full of *qualia*, such as seeing the colour blue, hearing a musical note or tasting watermelon. *Qualia* are central to consciousness.

CONSCIOUSNESS IS THE BRAIN?

If you ask philosophers "What is the nature of consciousness?" a range of very different answers will come back. There is no agreed theory. However, two views in particular receive a regular hearing among reductive physicalists (those who hold the view that the mind is reducible to physics and chemistry alone). One is that brain science can access consciousness and explain it entirely. Another is that consciousness is illusory. Both views ultimately believe that consciousness is synonymous with brain activity. In other words, consciousness *is* the brain.

Let's consider the first view. Is it true that scientific methods can access and explain consciousness? One front cover of *New Scientist* in 2016[28] was entitled "The Metaphysics Issue: How science answers philosophy's deepest questions". Some of those questions included "Why does anything exist?", "Can I know if God is real?", "What is time?", and "What is consciousness?" It was an ambitious title, implying that scientists are able to parachute in and answer questions that have long stumped the finest minds in philosophy. In reality, scientists face an impasse in terms of accessing consciousness itself. Yes, using fMRI we can see and measure brain activity as various thoughts happen. But this is a far cry from seeing someone's *actual* thoughts. How can a scientist access someone else's inner life? How can scientific methods access *qualia*?

Imagine a friend returns from a concert and is raving to you about it. The warm-up gig, the atmosphere, the lights, the crowds and their favourite songs all feature. Intrigued, you go away and read the reviews. These reviews capture some details, but not what it *would have been like for you to be there and experience it yourself.* You would have had to be there yourself to really understand.

When scientists study consciousness, they can only approach it as an observer, similar to reading reviews. But consciousness is experiential. In the end, your friend who went to the concert, after enthusing about it for 20 minutes, might notice your raised eyebrows and say, "I guess you had to be there". The scientific method offers third-person observations, whereas conscious experience is encountered in the first-person. We can find out what's in someone's brain by measuring chemicals and electrical activity and recording MRIs, but we can't measure what's in their mind in the same way. To find out what's in their mind we need to ask the person to share their inner world with us. Scientists may help us understand certain aspects and states of consciousness, but they cannot get inside someone's head and recreate their actual experience. They cannot access *conscious experience itself.*

"EASY" AND "HARD" PROBLEMS

David Chalmers, Professor of Philosophy at the Australian National University, makes a distinction between the "easy" and "hard" problems of consciousness. Easy problems are concerned with explaining some of the correlations between conscious experience and brain activity. For example, scientists may look at the areas of the brain involved when consciousness changes from one state to another, such as from wakefulness to sleep. The "easy" problems are by no means easy. Many lifetimes of research are being taken up by these quests. Yet, compared with the "hard" problem they are trivial.

The "hard" problem involves accounting for conscious experience. How do you get from brain cells firing to "what it is like to be you'? Baroness Susan Greenfield, Professor of Physiology at the University of Oxford,[29] gave a lecture at the University of Melbourne in 2012, on "The Neuroscience of Consciousness". A lecture with this title has a "reveal all" tone to it. People show up and tune in wondering if the "hard problem" has finally been cracked. It was an engaging and brilliant lecture all in all, yet from the very beginning Professor Greenfield made her intentions clear:

> *I should perhaps say from the outset, what we are not going to be able to do is work out how the water is turned into wine. How does the water of boring old brain cells and sludgy stuff translate into the wine of phenomenological subjective experience.*[30]

The lecture would be confined to "easy" problems because,

true to form, the hard problem remains exactly that: *hard*. Chalmers himself does not believe that brain science will solve the hard problem.[31] We will say more later about his distinctive attempts to solve the "hard" problem. But first, let's indulge ourselves in a thought experiment.

MARY'S ROOM

Mary is a scientist with a detailed knowledge of the physics and chemistry of human vision. She knows all about the structure of the eye: how the cornea at the front focuses incoming light onto the retina at the back. Mary knows all about how the retina consists of rods and cones that process monochrome and colour vision respectively. She knows all about how this generates an electrical signal that is sent to the brain through the optic nerve and is converted into images.

But is this knowledge enough to understand what it is like to *see*?

In this thought experiment, the problem is that Mary was born blind, and so, for a large part of her life, she understood the workings of the eye in great detail without herself having the ability to see. Yet one day, the thought experiment goes, Mary miraculously gains the ability to see.

Here is the important question: at the moment of receiving her sight, does Mary learn anything new about vision? The author of this thought experiment, Frank Jackson, makes the point that if the answer is "yes", then physical facts alone cannot explain the first-person experience.[32] No amount of knowledge of rods, cones, corneas, light transmission, electrical stimulation of the optic nerve and image generation in the brain would get Mary any closer to the experience of what it is like to actually *see*.

Consider the body's response to pain. Imagine you are cutting bread and you accidentally slice your finger. The cutting of the skin sends a message to receptors in the skin, called nociceptors, which send a message to your brain that a painful stimulus has occurred. At the same time, you *experience* the sensation of pain. The skin receptors are important, but they are not a complete explanation. They contribute to the experience, but they are not the experience in its entirety. Or, returning to the coffee shop one more time, knowledge of the chemical structure of caffeine or its impact on brain physiology is of little help in describing the smell of coffee.

The water of brain processes alone cannot explain the "wine" of the Guatemalan blend.

IDENTICAL MEANS INDISCERNIBLE

On this basis, we can conclude that consciousness cannot be synonymous with brain activity. The two may well work closely together, but they are not identical. Philosophers refer to something known as the "indiscernibility of identicals", which originates with Gottfried W. Leibniz.[33] The essential idea is that if two things are identical, there would be no discernible difference between them.

Some years ago, my husband and I attended our first parents' evening to find out what kind of start our eldest child had made at primary school. The teacher began by making a few brief remarks about our daughter Abby. It was clear that we were talking about the same person. The teacher's comments about Abby's core personality and qualities were true of our daughter, because our daughter and the pupil Abby Dirckx are identical. Of course, parents usually know their children better than the teacher, and some behaviours come out at school but not at home and vice versa. But if the teacher began describing a child very different from Abby, we would quickly realise that they have the wrong person: her pupil and our daughter would not be identical.

If consciousness were synonymous with brain activity, then the two would be identical on every level. Everything true of consciousness would be true of the brain as well. In fact, as shown in the last few pages, they could not be more different. Therefore, a reductive physicalist approach to consciousness *must* be false. Consciousness simply *cannot* be reducible to physical processes in the brain.

HOW THE MIND WORKS

In 2000, Eric Kandel won the Nobel Prize for Medicine for his discovery that the process of learning causes brain cells to connect and grow.[34] For hundreds of years, the brain had been thought of as a static machine that undergoes gradual decline. Kandel's work marked a key moment in thinking of the brain as "plastic". This has nothing to do with cling film or sealable containers, but rather, describes the brain's ability to change and make new connections in response to the mind and a person's activity.

It is a sobering fact that our grey matter can shrink and sever connections, or it can grow and make new ones. When we get better at a musical instrument, it is because new connections have been laid down. When we forget a detail or become rusty at golf, old connections have broken away. The brain is very responsive to the mind. Reductive physicalists would have us believe that the brain 'drives' the mind entirely. But brain plasticity makes it clear that we are not simply at the mercy of our neurons driving every thought and action. Our thoughts can also impact our neurons and their pathways. The mind is able to bring about changes in the brain, through a process known as *downward causation*.

Some cities have complicated one-way road systems. We know the frustration of getting lost in them, with no possibility of turning around to retrace our steps. How much easier it is when the road has lanes travelling in both directions. Turning around is more straightforward. Downward causation tells us that there is not simply one-way traffic from the brain to the mind; there are also "lanes" in the opposite direction, from the mind to the brain.

Consider the placebo effect, which is considered by many scientists and clinicians to be the power of the mind to heal the body.[35] If a patient is given a "blank" or placebo treatment that they believe will help, this can in fact bring some healing to the body. The brain is triggered to release endorphins, the body's natural painkillers, which bring relief. The placebo effect is an accepted phenomenon. It illustrates the power of the mind over the body and brain, suggesting that the brain cannot simply drive everything. Brain chemistry may influence conscious awareness, but conscious awareness can also alter the chemistry in our brains.

ALL IN YOUR HEAD?

Neurologists, such as Suzanne O'Sullivan, tell us that the power of the mind over the body can also lead to physical illness. Her book, *It's All in Your Head*, shares stories from 20 years in the clinic, of how some patients have dramatic symptoms but no detectable physical cause. Very often, this kind of illness has been shown to have an emotional cause, in which "physical symptoms ... mask emotional distress".[36] O'Sullivan writes:

> *Psychosomatic disorders are conditions in which a person suffers from significant physical symptoms— causing real distress and disability—out of proportion to that which can be explained by medical tests or physical examination. They are medical disorders like no others. They obey no rules. They can affect any part of the body.*[37]

In 1997, the World Health Organization conducted an investigation into psychosomatic illness and found that it affects 20% of patients worldwide, in countries such as the USA, Nigeria, Germany, Chile, Japan, Italy, Brazil and India. In 2005, the yearly cost of treating psychosomatic illness in the USA was estimated at $256 billion, compared with, say, diabetes, which costs $132 billion per year.[38]

The assertion that "you are your brain" is insufficient to explain this kind of illness—or to do anything about it. There is more to us than simply our brains. The mind is powerful in its effect on the body.

CONSCIOUSNESS OBJECTORS

Let's consider now the second view that consciousness does not exist at all. Several philosophers and neuroscientists take the view that there are no first-person experiences. Psychologist Susan Blackmore does not deny consciousness entirely but holds that there is no continuous stream of consciousness.[39] Tufts University Professor of Philosophy, Daniel Dennett, and others,[40] take the thinking further and hold that if consciousness is entirely physical, the perception of it being anything more is the result of "an amazing collection of mundane tricks in the brain".[41] Dennett's book *Consciousness Explained* draws these conclusions, and at his inaugural lecture at the Hungarian Academy of Sciences in 2002, he put it like this:

> *Consciousness is a physical, biological phenomenon ... that is exquisitely ingenious in its operation, but not miraculous or even, in the end, mysterious. Part of the problem of explaining consciousness is that there are powerful forces acting to make us think it is more marvelous than it actually is. In this it is like stage magic, a set of phenomena that exploit our gullibility, and even our desire to be fooled, bamboozled, awe-struck.*[42]

NO "HARD" PROBLEM?

According to Dennett, there is no "hard" problem. Dennett solves the hard problem by denying it. There is no first-person experience. There isn't "something" that it is like to be *you*. Consciousness doesn't exist. There is simply neuron-firing and brain chemistry. Any sense that there is more going on is illusory. What shall we make of this view? Is it true that our brains are fooling us into believing a false reality? If that is so, how do we know it isn't true of every thought that we have, including the very notion that consciousness is illusory?

Moreover, illusion still presupposes consciousness. Illusion occurs when an experience is misinterpreted or perceived wrongly. But the experience itself is still valid and real. When we watch a magician, their skill and sleight of hand lead to all kinds of surprises for the audience. The seemingly impossible becomes possible; things appear real to us when they are not. But we still *experience* the magic show, even if we misread what is happening.

As Dennett himself said, there is still an "us" that can be

"fooled, bamboozled, awe-struck". Consciousness undergirds even illusion.

Dennett seeks to eliminate the first-person reality, but he has not succeeded in eliminating the first-person from his vocabulary. Search for quotes by Dennett on Google, and you don't have to look far to find this line,

> There's nothing I like less than bad arguments for a view I hold dear.[43]

In what sense "I" if there is no first-person perspective? Dennett's view in the end is simply absurd and makes it impossible to say anything at all. Susan Blackmore acknowledges that chalking it all down to illusion raises more questions than it answers:

> If you say "it's all just an illusion" this gets you nowhere—except that a whole lot of other questions appear. Why should we all be victims of an illusion, instead of seeing things the way they really are? What sort of illusion is it anyway? Why is it like that and not some other way? Is it possible to see through the illusion? And if so, what happens next?[44]

Dennett's view undermines itself. Worse still, it undermines the very concept of rational thought. What are these "powerful forces acting to make us think [conscious experience] is more marvelous than it actually is"? How do we know they are not also at work in distorting Dennett's perspective? Dennett assumes he is outside of the concepts he seeks to explain, but if what he asserts is true then his very argument cannot be trusted. It backfires. Some have therefore renamed Dennett's book: "Consciousness Explained Away".[45]

Furthermore, on a very practical level, we do not and cannot live without assuming there is "something it is like to be *us*". Neuroscientist Christof Koch brings the point home very candidly,

> *If I have a tooth abscess ... a sophisticated argument to persuade me that my pain is delusional will not lessen its torment one iota.*[46]

We live as though our unique personal experience of life is real and to be taken seriously. We write autobiographies, and we read them, by and large assuming their truthfulness. We respond to emergency appeals for humanitarian aid assuming many thousands of people each have a unique experience of suffering. Even the "post-truth" society that we are told we now live in "defines truth" by experience. Yet, if there is no first-person experience then even the concept of being post-truth is meaningless. We have tied ourselves up in philosophical knots. If we want to deny consciousness then all kinds of other things will have to be written off as well. We cannot have our conscious cake and eat it.

In summary, there are persuasive reasons to discount the view that consciousness *is* the brain, from philosophy, neuroscience and medicine. But if we are not just machines, the question that naturally follows is are we are *more* than machines? If so, what might that look like? This will be the subject of our next chapter.

Are we more than machines?

I peered into Amy's eyes. All I saw was emptiness. That same deep well of emptiness that I had seen countless times before in people who, like Amy, were thought to be "awake but unaware". Amy gave nothing back. She yawned. A big open-mouthed yawn, followed by an almost mournful sigh as her head collapsed back onto the pillow.

Seven months after her accident, it was hard to imagine the person Amy must once have been—a smart college-varsity basketball player with everything to live for. She'd left a bar late one night with a group of friends. The boyfriend she'd walked out on earlier that evening was waiting. He shoved her and she toppled, slamming her head on a concrete curb. Another person might have walked away with a few stitches or a concussion, but Amy was not so lucky. Her brain hit the inside of her skull. It pulled from its moorings. Axons stretched and blood vessels tore as ripples of shock waves lacerated and bruised critical regions far from the point of impact. Now Amy had a feeding tube surgically inserted into her stomach that supplied her with essential fluids and nutrients. A catheter drained her urine. She had no control over her bowels, and she was in diapers.

Two male doctors breezed into the room. "What do you think?" said the more senior of the two, looking straight at me.

"I won't know unless we do the scans," I replied.

"Well, I'm not a betting man, but I'd say she's in a vegetative state!" He was upbeat, almost jovial.

I didn't respond.

The two doctors turned to Amy's parents, Bill and Agnes, who'd been patiently sitting while I observed her. A good-looking couple in their late forties, they were clearly exhausted. Agnes gripped Bill's hand as the doctors explained that Amy didn't understand speech or have memories, thoughts or feelings, and that she couldn't feel pleasure or pain. They gently reminded Bill and Agnes that she would need round-the-clock care for as long as she lived. In the absence of an advanced directive stating otherwise, shouldn't they consider taking Amy off life support and allowing her to die? After all, isn't that what she would have wanted?

Amy's parents weren't ready to take that step and signed a consent form to allow me to put her in an fMRI scanner and search for signs that some part of the Amy they loved was still there. ....

Five days later I walked back into Amy's room, where I found Bill and Agnes by her bedside. They looked up at me expectantly. I paused for a moment, took a deep breath, and the gave them the news that they hadn't allowed themselves to hope for:

"The scans have shown us that Amy is not in a vegetative state after all. In fact she's aware of everything."

After five days of intensive investigation we had found that Amy was more than just alive – she was entirely conscious. She had heard every conversation, recognised every visitor, and listened intently to every decision being made on her behalf. Yet she had been unable to move a muscle to tell the world, "I'm still here. I'm not dead yet!" [47]

Human consciousness and its relation to the brain is no straightforward matter. Professor Adrian Owen has spent more than 20 years studying patients with brain trauma, and his book *Into the Grey Zone* tells the story of several of his patients, including Amy mentioned above. For Owen, as a clinician, the question of central importance was this: is there any possibility that a patient who appears to be vegetative is, in fact, conscious? Using brain-imaging techniques, Owen's team at the University of Cambridge discovered that the answer to this question was *yes*. Their groundbreaking results were published in the prestigious journal *Science* in 2006. [48]

Of course, by 2006, locked-in-syndrome, was not a new concept. It came into the foreground in the 1990s, aided by the movie *The Diving Bell and The Butterfly*, which told the story of Jean-Dominique Bauby and of the road that he and his loved ones had walked in the aftermath of his massive stroke. Neurologists have since developed ways of testing whether or not a patient is awake, by enabling them to respond to questions using, for example, very small eye movements. Owen's team took this one stage further and developed ways in which an entirely immobile patient can respond with their mind.

Scientists discovered that a small number of patients considered vegetative as a result of severe brain trauma were able to follow instructions and answer questions about themselves using mental exercises. In other words, they were conscious despite having had significant brain damage. What are we to make of these discoveries in view of the assertion: "You are your brain"? Owen's study could be seen as one of several "signposts" from within neuroscience itself indicating that human consciousness surpasses the condition of our brains. Could it be that we are more than our brains and more than simply physical machines? Owen himself seems to think so, and summarises his research in this way:

We have discovered that 15 to 20 percent of people in the vegetative state who are assumed to have no more awareness than a head of broccoli are fully conscious, although they never respond to any form of external stimulation. They may open their eyes, grunt and groan, occasionally uttering isolated words. Like zombies, they appear to live entirely in their own world, devoid of thoughts or feelings. Many really are as oblivious and incapable of thought as their doctors believe. But a sizable number are experiencing something quite different: intact minds adrift deep within damaged bodies and brains.[49]

In this chapter, we will look at three attempts to explain why human beings are more than machines. We will first consider the view that the brain generates consciousness, second, the belief that consciousness resides in every living thing, and third, the possibility that consciousness is beyond the brain.

DOES THE BRAIN GENERATE CONSCIOUSNESS?

Phineas Gage was a railway worker in 19th-century Vermont. The method for clearing rocks at that time was to drill holes, push dynamite into them with a metal rod, light the fuse and run. In 1848, when Gage was 25, the rod he used to push down the dynamite made a spark which ignited the dynamite prematurely. The metre-long rod shot through Gage's skull and brain and landed some 30 metres away, destroying most of his left frontal lobe. Gage was presumed dead, yet he regained consciousness after a few minutes and went on to live for another twelve years. The story goes that he underwent a dramatic personality change. This previously mild-mannered man became uninhibited and prone to profanity and inappropriate behaviour, so much so that he lost his job.

The medical literature today is full of reports of personality changes that follow a brain injury, similar to that of the famous case of Phineas Gage. We see this all too prominently in those with degenerative diseases such as Alzheimer's disease and dementia. A key feature of these illnesses is that damage to the brain has a profound effect on the mind. As the brain unravels, the person's memory, power of recall and, often, personality, follow suit. A number of brain syndromes also impact the mind and conscious experience of the person: for example, Cotard's syndrome, in which the patient believes they don't exist.[50] Some epilepsy patients are treated by a surgical procedure called a callosotomy, which disconnects the left and right sides of the brain. As a result, a proportion of patients report being aware of two "selves".[51] Whenever the brain is damaged or injured, the mind similarly suffers. Even hitting someone over the head with a blunt object will, at the very least, give them a headache and impair their ability to think clearly. In healthy brain development, the same parallels are seen. The level of consciousness of a newborn baby is different from a toddler, which is different again from a teenager, because their brains are at different stages of growth. Mind and brain—consciousness and brain—though not identical, are clearly connected.

Many agnostics, atheists[52] and Christians[53] make sense of this close connection between mind and brain by taking the view that the brain *generates* mind and consciousness.[54] When a number of different parts come together over time, a new thing comes into being, but if those components are broken up, the new entity goes away. This view is broadly known as non-reductive physicalism, outlined in chapter 1.

Consider the case of a university. A university as an institution is made up of many different departments, each with its own subject area and expertise. The university arises out of the fact that there are many different departments generating research. But a university is more than the sum of its departments. A university also has an alumni network, an international reputation, a donor base. Members of that university develop ideas that shape culture and make discoveries that can change the course of history. In the case of a university, the institution is formed by its component parts but is far greater than all of them combined. Yet if those departments were all dismantled, the university would cease to exist.

Similarly, according to this view, mind and consciousness emerge through the coming together of a number of physical parts, but are more than the brain alone. So, what are we to make of this view?

THE PROBLEM IS STILL "HARD"

Museums are fascinating places. Collections of ancient artefacts encased in glass evoke a variety of reactions in both adults and children. We gaze at ancient pieces of wood, stone and metal, trying to comprehend the years that have passed since that spear or pot was in the hands of the person who actually made it. Ancient artwork and jewellery are particularly fascinating. In anthropological terms, expressing oneself through creativity is a sign of advanced thinking, abstract thought, and, ultimately, human levels of consciousness. This kind of expression was initially absent, yet became integral to the life of homo sapiens during the late Stone Age.[55]

The question is, what brought about such a change? The concept of *emergence* is offered as the natural solution to the "hard" problem mentioned earlier. Yet, it actually raises more questions. *How* exactly did consciousness emerge? If we are dealing with a closed system of meaningless matter and non-conscious neurons, how did these come to generate conscious minds? The hard problem has not gone away. Even though emergence makes sense in some respects, there is still a vast area of unknowns for which there are no clear answers.

Further questions arise. If the only basic ingredient needed to generate consciousness is the physical brain, why do animals not have access to the same levels of consciousness as humans? Genetically speaking, humans and chimpanzees share a significant amount of DNA. Primates seem to possess some conscious awareness, but they are not conscious to the same extent that humans are.[56]

Take the dog, for example—man's best friend. Dogs are highly tuned to their environment. They cower during a thunderstorm. They bark and whine when their owner is out. They chase rabbits and other dogs. They even experience emotions: a fearful dog has its tail between its legs, whereas a happy dog wags its tail vigorously. But as far as we can tell, they do not appear to have an inner life. They do not crawl into their bed thinking, "I had a difficult day today". They do not ask themselves, "Why am I a dog and not a cat?" or "What is the purpose of my life?" Dogs eat, sleep and respond to their environment.

Harvard psychologist Steven Pinker argues that primates can be trained to use language and perform cognitive tasks in a limited way. True. But only if a human initiates the training! Max Tegmark, writing in the *New Scientist* book *The Universe Next Door*, is among a number of philosophers who explain the differences in terms of complexity. When groups of atoms are arranged in new ways, new properties emerge.[57] Higher and higher levels of complexity lead to more and more sophisticated abilities. However, studies show that even the most developed and highly-trained primate cannot exceed the cognitive ability of a 4 year-old child.[58] The discontinuity between apes and humans is one of "kind" and not "degree". Complexity alone is insufficient to get us across the chasm.

Philosopher J.P. Moreland makes the point that, in a material-only world, the view that the brain generates consciousness is vulnerable to more persuasive rival theories.[59] We will say more later about what kind of rival theories those might be.

WHAT IF THE BRAIN DIES?

In 1991, Pamela Reynolds suffered a severe brain haemorrhage caused by an aneurysm, requiring life-saving surgery. The operation, nicknamed "Stand still" needed her body temperature to be cooled, her heart and brain signal to be "flatlined", and the blood drained from her head. Technically speaking, Pamela was clinically dead. The operation was successful, but when resuscitated, to the surprise of her doctors, Pamela recalled being conscious during the surgery.

Cardiologist Dr Michael Sabom recounts Pamela's story in his book *Light and Death,*[60] in which he looks at the science of Near Death Experiences (NDEs). Several decades later, as cardiac surgery has increased, stories like Pamela's are frequent. Hundreds of people have described suddenly being out of their body, looking down on the room with visions of white tunnels and deceased relatives combined with a feeling of peacefulness and wellbeing. Some have also described more distressing NDEs.

A Gallup survey conducted in 1982 across millions of ordinary people revealed that 15% have had an NDE. Since so many people have reported them in the last three decades, several cardiologists,[61] psychologists[62] and paediatricians[63] from different continents have systematically interviewed thousands of patients and tried to understand this phenomenon. Books have been written by distinguished surgeons,[64] scientists[65] and philosophers alike.[66]

NDEs are a key objection raised against the view that the brain generates consciousness. If consciousness is contingent on the brain, then, when the brain dies, we would expect consciousness to cease. But are NDEs telling us that you can be conscious without a functioning brain? Or are can they be explained in other ways?

OBJECTIONS

There are concerns that NDEs may be fabricated, caused by residual brain activity,[67] or merely be wishful thinking. We cannot rule out the possibility of fabrication; however some patients have recounted details from their near-death state that could be externally corroborated: details such as the particulars of a conversation taking place several rooms away at the time of surgery,[68] or those blind since birth claiming sight and describing family and friends in detail.[69]

As for residual brain activity, we can't rule out the possibility of neuronal activity that our instruments cannot detect. But the patient is in a state of clinical death. The blood supply to their heart and brain has been temporarily ceased to enable surgery. Even though clinical death is reversible and differs from full biological death, the notion of any residual brain signal in this state is hard to account for.* [70] Any wishful thinking by those who are scared to die does not seem to explain NDEs either. If the person is clinically dead, then surely fears of dying are irrelevant. Furthermore, some NDEs are frightening or happen to those with no beliefs about heaven or hell.[71]

Do NDEs call into question the view that consciousness is contingent on the brain? Some say "yes". NDEs may seem bizarre and contrary to ordinary experience. Yet, even if only one authentic encounter of this kind has ever happened, it is a serious blow to the view that consciousness resides entirely in, or even emerges from, the physical brain. Others say "no" and regard this data as speculative and in its early stages. For Christians, if NDEs are real, they do not suggest a one-size-fits-all truth to the afterlife. They are describing

---

* In a study of 1,400 NDEs over 18 years by cardiologist Fred Schoonmaker, fifty-five patients had NDEs when there was no recordable EEG signal.

what happens as a person *approaches* death or is in the very first few moments of reversible clinical death, which is different from full, irreversible biological death.

PROOF OF HEAVEN?

I arrived at university in 1993 as an agnostic and quickly began to meet people from a variety of different backgrounds and beliefs. I vividly remember a conversation about life after death with a friend who was an atheist. His opinion? When you die: oblivion. *End of story.* I didn't feel strongly either way.

It wasn't until later that I found myself asking more questions. *What if God exists anyway?* We are all entitled to an opinion, but opinions become irrelevant when they meet facts. What if God is real? What if it is a fact that I will meet him one day and have to give an account of my life? Will I be ready? Will my excuses for how I have lived stand up? Questions about life beyond death were a key part of my journey towards faith.

Are NDEs proof that heaven is real? The prospect of consciousness beyond death is attractive to some but sobering to others. Some speak of their loved ones as being in a better place, as though they do continue to exist. Others are fascinated with trying to contact the dead through séances, therefore assuming that the person still exists albeit, "on the other side". Still others have no desire to live for ever. But are NDEs proof that there is a "better place" or even that God exists?

Christians should be hesitant to posit NDEs as proof of God, although Jewish neurosurgeon Eben Alexander has put it more strongly and has referred to them as "proof of heaven".[72] This field continues to develop—however, if

God exists, then the possibility of consciousness beyond the death of the brain should come as no surprise. One of the most famous verses in the Bible tells us that...

> *God so loved the world that he gave his only Son, that whoever believes in him shall not perish but have eternal life.* JOHN 3 V 16

The claim is that death need not be the end. There is more that awaits us. There is the possibility of life and consciousness beyond the grave.

IS EVERYTHING CONSCIOUS?

Some philosophers acknowledge the shortcomings of trying to explain consciousness in terms of matter alone, and recognise the need to start the conversation somewhere new. One such person is Professor David Chalmers, mentioned in chapter 3. According to Chalmers, the building blocks of the brain are not the right starting point. We need to start with our subjective experience, take this to be fundamental, and try to build a theory of consciousness around experience.[73] Chalmers solves the hard problem by stating that there is only one kind of substance, but it is both physical and conscious in nature. In this view, *everything* is conscious to some extent. All particles have both physical and conscious dimensions. More complex systems possess higher levels of consciousness but conscious states are present in every atom, including inanimate things such as minerals, metal and mould.

This view is known as panpsychism, derived from the Greek words, *pan,* meaning "all', and *psyche,* meaning "soul" or "mind". Panpsychism also underlies Buddhist and Jainist thinking from the East. Though intriguing, this view is difficult to visualise in real terms. Is it true that electrons, rocks, and trees each have some degree of consciousness? If so, what does this actually mean in practice? This view is impossible to verify. Nevertheless, it is popular among philosophers today because it solves the enigma of consciousness by claiming its presence in every aspect of life.

Another naturalist who rejects a strictly physicalist approach to consciousness is Thomas Nagel, Professor of Philosophy at New York University. Nagel takes a slightly different view to Chalmers and believes that atoms and molecules are precursors to consciousness: a view known as protopanpsychism. According to this view, the quality of the first-person experience increases as complexity increases from atoms to cells to organs and organisms all the way up to human beings. An electron has a different experience of the world than a human, but the potential for consciousness exists nonetheless.

Critics of protopanpsychism argue that it does not really solve the hard problem so much as raise new questions. *What does it mean to be pre-conscious?* As philosopher William Jaworski puts it:

> *We know what beliefs or desires are, but what are protobeliefs and protodesires? I believe that 2+2=4, for instance, but what would it mean to have a protobelief that 2+2=4?* [74]

Nevertheless, Nagel has stuck his head above the parapet, and in his most recent book, *Mind and Cosmos,* suggested that...

> *we should not renounce the aim of finding an integrated naturalistic explanation of a new kind.* [75]

His views and their expression have come at some cost to

him in the physicalist community. Nagel has been branded a "heretic" by some for daring to question the view that matter explains everything.

So, one possible rival theory is the belief that matter itself possesses consciousness. Consciousness is basic to life. What other possible theories are there?

IS CONSCIOUSNESS *BEYOND* THE BRAIN?

A third view is that consciousness is *beyond* the brain. Conscious experience is a fundamental and basic building block of life, and we need to begin here and explain everything else in relation to consciousness, rather than the other way around. Conscious states are independent of neurons and brain chemistry. In fact, there are two distinct, but interactive substances at play: a physical brain and a non-physical mind that is conscious.

Many scientists are quick to dismiss this view, known as substance dualism. A *New Scientist* article entitled, "What is Consciousness?" argued that…

> *[Descarte's] ideas [about substance dualism] influenced neuroscience until a few decades ago, but the field has moved on.*[76]

Not according to a number of eminent thinkers. Several professors of philosophy continue to be substance dualists today including Richard Swinburne[77] and Keith Ward[78] at the University of Oxford, Alvin Plantinga at Notre Dame[79] and J.P. Moreland of Biola University[80].

Michael Egnor, Professor of Neurosurgery at Stony Brook University School of Medicine, New York, makes the point that many of his patients are missing large parts of their brains, yet have "quite good minds". A person can remain intact despite significant damage to the brain. He recalls even being able to hold a coherent conversation with a patient while removing a tumour from her frontal lobe. After 30 years of working in this field, Egnor has concluded that the mind of a person is beyond simply the workings of their brain, and comments that...

> *materialism, the view that matter is all that exists, is the premise of much contemporary thinking about what a human being is. Yet, evidence from the laboratory, operating room, and clinical experience points to a less fashionable conclusion: Human beings straddle the material and immaterial realms.*[81]

Even patients with advanced dementia still have moments of lucidity. Some callosotomy patients may speak of having "two selves" but not of having "no self". Neurosurgeon Wilder Penfield (1891-1976), is referred to by some as the father of modern neurosurgery. His groundbreaking studies on conscious epilepsy patients in the 1950s enabled the function of many regions of the brain surface to be identified. But Penfield never stimulated an area that changed or induced the person's sense of self.[82] All kinds of involuntary sensations and movements were triggered—sometimes even emotions—but never abstract reasoning. Never a sense of what it is like to be you. Never consciousness itself. It seems as though the field of neuroscience has not moved on from substance dualism. A number of clinicians believe a non-physical mind makes the best sense of their observations.

Even the story of Phineas Gage does not rule out the possibility that consciousness is beyond the brain. His case-study is often used as evidence that the mind is entirely dependent on the brain. However, Marilynne Robinson, in her book *Absence of Mind,* makes the point that although the story of Gage has likely been embellished over time, he did work again,[83] and anyone who has survived a head-on collision with a metal rod might be forgiven for being a little upset! Perhaps this personality change cannot be pinned entirely on the trauma to Gage's brain. The story of Phineas Gage may even make the opposite point from that which it is often intended: even with the most severe damage to the brain, the person can continue to function.

HOW DO BRAIN AND MIND INTERACT?

The question of how a non-physical mind could exert changes in a physical brain poses concerns for many. If mind and brain are distinct, how do we explain the clear interaction between them? Surely Descartes' "ghost in the machine" implies that changes to the brain ought to have no effect on the mind? But this is clearly not the case.

Dualists respond that human mind/brain relationship is not confined just to Cartesian dualism in which mind and brain interact only through the pineal gland (see Chapter 2). Many dualists today are more holistic in their approach and favour models such as Thomistic dualism—that is the view that conscious states exist beyond the brain but are also causally connected to the brain. A holistic dualist accepts and welcomes the discoveries of modern neuroscience but would add that they are not the whole story.

Further, the non-physical can change the physical in other areas of life. Cyber-bullying can cause a child to lose their appetite, have panic attacks and lose sleep. Being asked on a date by the love of your life may cause blushing and incoherent speech but at the same time put a spring in your step. Crying, a process that releases salt water from the lacrimal glands in our eyes, could be triggered by the news that a loved one has died or that a friend has passed their exams. Information of this sort is non-physical, but it has a physical effect. The immaterial impacts the material in daily life, all the time. So why not a non-physical mind interacting with a physical brain?

UNSCIENTIFIC?

Some are quick to dismiss the possibility of a non-physical mind on the basis that it is unscientific. Yet there are some scientists who take this view and other scientists who don't. There are even scientists who have changed their minds. Harvard Biologist Professor George Wald, Nobel Prize-winner for his work on the biochemistry of vision, began life as a staunch atheist but underwent a dramatic change of beliefs in his seventies. He later reached these conclusions about consciousness...

> *Consciousness seems to me to be wholly impervious to science. It does not lie as an indigestible element within science, but just the opposite: Science is the highly digestible element within consciousness ... Mind, rather than emerging as a late outgrowth in the evolution of life, has existed always as the matrix, the source and condition of physical reality—that the stuff of which physical reality is composed is mind-stuff.*[84]

AN OPEN SYSTEM?

A decision about the nature of consciousness cannot ultimately be reached on the basis of science. It really comes down to a personal worldview. What if we considered the possibility that we don't live in a closed system of meaningless matter? What if it turns out that there is meaning in the universe as well? What if we entertain the possibility that God exists? How would this help us with the "hard" problem? It would mean that we can consider consciousness to be fundamental to the universe once again, but in a different way to our panpsychist colleagues.

Christians who are non-reductive physicalists take the view that the brain has given rise to the conscious mind but as the creative handiwork of a conscious being—God. In this view, the bridge to human consciousness is not traversed by greater and greater levels of brain complexity, but by humanity entering into a relationship with their Maker.[85]

The very first chapters of the Bible poetically and creatively describe the formation of human beings,

> *The Lord God formed the man from the dust of the ground and breathed into his nostrils the breath of life, and the man became a living being.* GENESIS 2 v 7

These verses are not necessarily at odds with scientific descriptions of the *processes* by which homo sapiens came to exist. But they imply that physical descriptions alone are not enough to describe the human person. The Hebrew word for "breath of life" is *neshama* or *ruach,* and means "God's breath" or "God's Spirit". According to these verses, a person is far more than matter. Far more than a machine. They have been breathed into by God, and it is this that has given them a capacity to think about themselves and beyond themselves to other people and to God himself. Other verses describe this in terms of humans being made "in the image of God" (Genesis 1 v 27). We will say more about what this might mean in chapter 8; suffice it to say that if God himself is conscious, we might expect humans to be as well.

MIND-STUFF

Christians who are substance dualists believe that finite, ir-reducible consciousness exists, and therefore there is a good chance that a conscious being known as God also exists. On one occasion, J.P. Moreland described the argument from consciousness to the existence of God in interesting terms, perhaps responding to the earlier language of "stage magic" from Daniel Dennett:

> We all know that you can't really pull a rabbit out of a hat. When a magician claims to pull a rabbit out of the hat, we know that there had to be a rabbit in the hat to begin with… Because you cannot get something out of nothing; you can't have nothing there and then all of a sudden a rabbit appears. If you start with matter from the Big Bang and all you do is rearrange it, according to the laws of chemistry and physics, you are not going to be able to get a conscious rabbit out of that mate-rial hat. You will end up with a very complicated hat but there will not be a rabbit. The reason conscious-ness exists is because we started with a rabbit, that is a great big rabbit—namely God—who is himself con-scious, and we do not have to pull a rabbit out of an empty hat and explain how you get consciousness from matter because there never was such a thing as just pure matter. God always existed.[86]

In other words, consciousness exists because God exists. We are conscious because God is conscious. God is a thinking, feeling, conscious being who is also relational and wants to extend consciousness beyond himself, to the people he has made.[87] Consciousness interacts with the brain in the case of humans but is not dependent on the brain in the case of

God. If God exists, then it is possible to be conscious without a brain. Contrary to popular opinion, the Bible is still a best-seller, and its very first sentence says:

> *In the beginning God created the heavens and the earth.* GENESIS I V I

"In the beginning God…" The "mind" that George Wald spoke about is the mind of God. According to this view, the mind of God has always existed and gave rise to everything else. If God exists, then the system is not closed, and there is hope for solving the hard problem.

But the question of whether we are machines or not leads on to another question. If the brain drives everything, then are we really free to make our own choices? Or do we simply do what our brains tell us? This will be the subject of our next chapter.

Is free will
an illusion?

You are out shopping for nibbles to bring to the office party and decide on breadsticks and hummus. Did you make the decision or did your brain make it for you? The party is great. You decide to take a taxi home instead of the bus. Did you make the decision or did your brain make it for you? The next day your boss announces cutbacks, and you have to decide whether to go forward for voluntary redundancy or not. You decide yes. Did you make the decision or did your brain make it for you?

We make decisions all the time. Mundane decisions like what to eat for breakfast. Important decisions like what courses to study or which city or country to live in. Sobering yet exciting decisions like whether or not to propose marriage. Our decisions mean something. Sometimes we get it right and reap the rewards. Sometimes we make the wrong decision and end up paying for it. But either way, the decision appears real and meaningful to us.

But if we are just our brains, are we really free in any meaningful sense? A number of scientists and philosophers strongly believe not. It may seem to us that our choices are freely made and therefore meaningful, but free will is actually an illusion, they argue. We are simply doing what our brains tell us.

The brain-imaging scientist and atheist Sam Harris began his book *Free Will* by describing a sordid and brutal attack on a family in Connecticut. It was an attack by two men, one of whom was called Komisarjevsky, that was only intended to be robbery, yet ended with four counts of murder. According to Harris, the attackers could not have done things any differently, even if they had wanted to. The violence inflicted on July 23, 2007 was a product of their genes and upbringing over the long term, coupled with their brain activity in the moment. Harris reflects:

> *As sickening as I find their behaviour, I have to admit that if I were to trade places with one of these men, atom for atom, I would be him … If I had been in Komisarjevsky's shoes on July 23, 2007—that is, if I had his genes and life experience and an identical brain (or soul) in an identical state—I would have acted exactly as he did. There is simply no intellectual position from which to deny this. The role of luck, therefore, appears decisive … How can we make sense of our lives, and hold people accountable for their choices, given the unconscious origins of our conscious minds? … Free will is an illusion. Our wills are simply not of our own making. Thoughts and intentions emerge from the background causes of which we are unaware and over which we exert no conscious control. We do not have the freedom we think we have.* [88]

It is disturbing to hear a respected scientist make a dogmatic statement that implies we cannot ascribe any meaning to what he is writing! Do science and philosophy necessarily take us to this place? Is something or someone making decisions for us? Is it true that there is no other intellectual position? Or is the case for free will still on the table?

HARD-DETERMINISM

Those, such as Harris, who answer this last question with a very strong no, are considered to be "hard determinists". Determinism is the belief that prior causes guarantee *a particular outcome*.[89] The free-will debate zooms in on one aspect of determinism, and asks: "Do prior causes guarantee *human decisions?*"[90]

A hard determinist believes that the human brain and the choices arising from it are entirely determined by prior causes. The human brain is akin to a machine that operates according to fixed processes, and this fixed nature of the brain rules out the possibility of free will.

Harris held that Komisarjevsky's behaviour was determined by several prior causes: his upbringing (social determinism), his genes (genetic determinism), and his brain activity on the day (neuronal determinism). Hard determinists are committed to believing that if Harris had been "in Komisarjevsky's shoes", he would have acted in exactly the same way. Komisarjevsky was locked into his actions and could not have behaved differently. For hard determinists, if the prior conditions are the same, even for two different people, the outcome never changes. Determinism and free will are incompatible, hence hard determinists are also known as incompatibilists. The 20th-century philosopher Friedrich Nietzsche held this view.

Is Harris correct that this is the only intellectually credible position? In short, no. There are also libertarians and compatibilists. According to these two views, the human will is not illusory. It is part of the mind and therefore able to bring about effects on the body. Libertarians argue that we always have the ability to choose freely and without constraint, whereas compatibilists argue that we are largely determined but can act freely under certain conditions.

COMPATIBILISM: "SOFT" DETERMINISM

"Compatibilism" or "soft determinism" holds that determinism is true, but is also compatible with free will. Compatibilists believe that human behaviour is determined by prior causes, but we can also act freely when we are not being constrained or are seeking to fulfil our desires. Think of the methods involved in flying a plane. The autopilot mode determines most of the details of the flight. However, in normal circumstances, the pilot is still able to make decisions from the cockpit, without coercion, that fulfil his desires for the flight: decisions such as the speed of the plane, the temperature of the cabin, and the general well-being of the passengers. Both automated processes and free decision-making work together and are compatible.

Our arrival on earth is clearly determined by prior causes involving our parents. Our genetic code is determined by the genes that they pass on to us, which in turn come from their parents and grandparents and so on. When it comes to the human brain, we cannot deny some level of determinism. Genetics and upbringing have shaped the brain each of us has today. Every brain is determined in that all have the same anatomy and the same networks for vision, movement, hearing, cognition and so on. Compatibilists emphasise that this is a good thing, and believe that human freedom is possible *because* of determinism in the brain, not *in spite of* it. Without automation, modern aeroplanes would not function regardless of the wishes of the pilot. The fixed properties of the brain provide structure and order, out of which we have the capacity to act freely under certain conditions.[91]

The 19th-century philosopher and atheist David Hume held this view, as does Tufts philosopher Daniel Dennett, today. A number of Christians are also compatibilists and believe that God arranges and determines the events of life, while still enabling certain freedoms.

LIBERTARIANISM

A third position held by philosophers is libertarianism. Its proponents argue that if determinism is true, the will cannot be free. In other words, determinism and free will cannot co-exist, and therefore libertarianism is also a second type of incompatibilism.

Libertarians believe that human freedom means freedom from all constraints. Sam Harris would not be doomed to repeat Komisarjevsky's actions because there is no determinism, hard or soft, that drives our choices. Any influences from our genes, upbringing or surroundings can be bypassed because the person is the true source of their actions. Brains don't make choices: people make choices. The will resides in the mind of the person, who always has the ability to act differently or, as philosophers say, bring about "alternative possibilities", if they so wish.[92]

There are several ways in which philosophers and scientists argue for libertarian free will, which we will focus on in the sections below.

QUANTUM DECISIONS

One approach that some Libertarians appeal to is the unpredictability of the brain at the atomic scale. The brain is not as locked in to particular patterns or responses as we might think and may have properties that explain free will. This is described by Heisenberg's uncertainty principle.

The theory goes that the location of larger brain entities such as brain tissue, blood vessels and ventricles can be known with certainty, but at the quantum level there is much more uncertainty. As the late Peter Clarke put it, "Nature is fundamentally fuzzy".[93]

Some libertarians—for example, the late Sir John Eccles— say that these perturbations are directed by the mind, and see Heisenberg uncertainty as a way of describing how the mind impacts brain function without breaking any physical laws.[94] Others view them as entirely random and too small to exert any significant changes in biological systems. They argue that Heisenberg uncertainty does not revive the concept of human conscious freedom; it delivers us chance activity.

Opponents of libertarianism caution that an indeterminate, chaotically functioning brain is just as much a threat to free will as is a fixed and determined brain.

TWO-WAY TRAFFIC

A second concept used by libertarians is downward causation. As we discussed in chapter 4 (pages 51-52), the brain is known to be plastic and responsive to its environment. The brain doesn't just impact the mind; the mind also impacts the brain. Downward causation tells us that there is not simply one-way traffic from the brain to the mind; there are also "lanes" in the opposite direction, from the mind to the brain. And if this is true, then we cannot simply be at the mercy of our neurons.

FREE AGENTS

The world record for the largest line of dominoes was broken again in April 2017. At the time of writing the current world record is well over 5 million. It is mesmerising to watch the knock-on effect as the pieces cascade in just a few minutes. Most of the conversation surrounding free will is taken up with how brain events lead to mental events, how neurons impact thoughts, or, in other words, how one domino causes the downfall of its neighbour. For a physicalist, dominoes are all that exist, and therefore explanations are restricted to this realm.

However, anyone who has seen a domino cascade will know that there is one action, crucial to its success, that was not caused by a domino. The finger movement of a person set it all going. In other words, there are different types of causation: domino-to-domino and person-to-domino. Professor Richard Swinburne, makes a distinction between "non-intentional causation", in which one event causes another, and "intentional causation", in which the cause is more than simply another event.[95] In fact, the cause is a distinct person: a free agent with a specific intention in mind who can exert causal power within the mind and brain. If this view, known as "agent-causation", is correct, then genuine libertarian freedom remains viable. Our brains do not make decisions; *we* make them as volitional human beings.

SCIENTIFIC ATTEMPTS TO UNDERMINE FREE WILL

Philosophers have debated free will for many centuries. However, the rise of modern neuroscience has led to recent attempts to undermine free will in the laboratory. A particular brain-imaging study, conducted in 1983 by physiologist Benjamin Libet, has had a pivotal impact on this question.

Libet took healthy volunteers and connected them to a machine that would measure brain activity through signals given on the surface of the scalp, a technique known as electroencephalography (EEG).[96] Libet asked volunteers to sit comfortably and simply move the index finger of their right hand at will. Participants were instructed to move their finger at a time of their choosing, and then note the time when they had consciously decided to do so.

BENJAMIN LIBET'S STUDY OF CONSCIOUS DECISION-MAKING

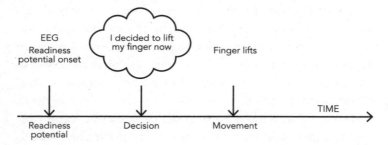

The results were controversial. They suggested that the brain started firing not only before the finger movement but also even *before* the person consciously chose to move it. Libet concluded that the brain decides to act long before the conscious mind of the person does. The brain drives the person. Similar brain-imaging studies followed on the heels of Libet's study, and discussion was sparked again among scientists and philosophers as to whether they now had scientific evidence that free will is illusory.[97]

RESPONDING TO LIBET

How should we make sense of Libet's data? Some follow-up analysis brought interesting results. Libet's initial data had been averaged across all 15 volunteers. But when broken down into individual people, the decision of a person preceded any brain activity in 5 out of the 15 volunteers. Furthermore, reanalysis of the data using a different method showed that the decision to move and the firing of the brain occurred at the same time.[98] Some have reinterpreted any initial spike as a sign that the brain is *ready* to make a decision, rather than representing the decision itself.[99]

Further studies also revealed that clock-monitoring explained some of the early brain-firing.[100] The volunteers had to watch a clock face in order to record the timing of their decision to move, but this had not been factored into the analysis. Later studies, with no clock-watching involved, showed little or no early activity.

But even if early brain-firing was confirmed, would one study of finger movement apply to all areas of decision-making? We make all kinds of decisions in life. How and when to bend a finger is the most basic of all. Can we really use these results to draw conclusions about all types of decision-making, including moral (or immoral) decisions—like choosing to murder someone?

Those who took part in Libet's original experiment made the decision to participate in the study hours, and probably days, before it began, rather than milliseconds before. Getting there would have involved decisions about transport, decisions about arranging cover at work, and decisions about whether the benefits would outweigh the risks. Once all of these decisions are considered, the impulse to move a finger seems trivial by comparison.

Neurosurgeon Wilder Penfield agreed that the electrical activity of the brain cannot explain free will. In his studies, in which he electrically stimulated areas of the brains of conscious patients in the 1950s, Penfield was able to cause patients to move their limbs and their mouths, and even experience memories and smells. But one thing he was not able to do was cause the patient to make a decision. He records, *"There is no place ... where electrical stimulation will cause a patient to believe or decide".*[101] Wilder started out as a materialist but his research left him with no choice but to conclude that some activities of the mind cannot be accounted for by the brain.

FREE "WON'T"

Even if impulses do arise in the brain without our awareness, we still have the capacity to allow them or stop them, and so we cannot simply be at the mercy of these impulses. The scientific term for this is "inhibitory control": our decision-making processes have control mechanisms built in. Everyday terms for this are self-control or resisting temptation.

Speaking personally, I find the impulse to eat biscuits arises most afternoons, but I have the choice of whether to give in to this or to eat the satsumas that wait patiently on my desk. My children occasionally have the impulse to whack their sibling over the head, but they have a choice of whether to follow through or to defuse the situation another way, ideally using words. Even in extreme situations this still applies. The criminal who is mid-robbery may have the impulse to kill, but, regardless of their brain or background, they still have a choice as to whether to give this impulse full vent or resist it.

Philosopher Michael Shermer describes this as "free won't". In Shermer's words:

> *Free won't is veto power over innumerable neural impulses tempting us to act in one way, such that our decision to act in another way is a real choice.*[102]

Libet published another paper in 1999 highlighting that he does, in fact, believe in free will precisely because of free won't. He concluded that…

> *the role of conscious free will would be … not to initiate a voluntary act, but rather to control whether the act takes place.*[103]

IS HARD-DETERMINISM COHERENT?

Sam Harris argues that "there is simply no intellectual position from which to deny [hard-determinism]".[104] Yet, if we look more closely, this view raises many concerns. In chapter 1, we outlined some helpful questions to test a belief or worldview: "Is it internally coherent?" "Is it externally relevant?" and "Can it be lived?" Let's ask these questions of hard-determinism .

Is it internally coherent?

Does hard-determinism make sense according to its own frames of reference? Not really. Hard-determinism makes it difficult to assert any personally-held belief. Religious people are sometimes accused of being programmed to believe in God. But, we could equally make the case that an atheist has been programmed to reject God, or an agnostic has been programmed to sit on the fence. Hard-determinism shows no favouritism towards one belief over another. It makes personally-held beliefs difficult to justify, and impossible to critique—including hard-determinism!

Hard-determinism also undermines any basis for rational discussion, as it impedes our ability to reason. If all thoughts are driven by non-rational, mechanistic forces it does not necessarily follow that the output of our minds will be rational. In other words, hard-determinism brings human rationality under fire, which in turn makes hard-determinism incoherent. It fails the first test.

Does it have explanatory power?

Does hard-determinism make sense of the world around us? Not really. For example, hard-determinism does not make sense of the fact that we humans strive for autonomy. Some secularists tell us that free will is an illusion, but if that is true, why do we continue to imagine it is real? So much of life centres around staying in control of things—our finances, our body shape, the demands of work. We want to make our own rules, decide our own fate, shape our own lives. So much fear surrounds loss of control over the things we hold dear. But why strive for autonomy if human freedom is an illusion? There is a deep contradiction here. Are we free to make our own rules or not? Or are even the rules we make for ourselves determined and therefore irrelevant? We cannot have it both ways. In terms of explanatory power, hard-determinism creates more confusion than clarity. It fails the second test. It determinism fails the second test.

Can it be lived?

Can hard-determinism be authentically lived out and lined up with our experience of life? Not really. We live as though our choices mean something: as though those decisions are made by us as volitional people, and not by the mechanistic firing of neurons in our brains. We are considered morally responsible and therefore accountable for our actions, good and bad. Hard-determinism threatens to completely unravel moral responsibility. Komisarjevsky was tried and sentenced to six life terms without possibility of release. Yet, if his actions of 23rd July 2007 were merely driven by forces beyond his control, then why punish him at all? Compatibilists or "soft" determinists have gone to great lengths to argue that even if the perpetrator could not have done otherwise, they are still morally responsible because they have acted in accordance with their own desires. However, if hard-determinism is true, then moral responsibility and the necessity of justice is removed, and the structures that keep ourselves as individuals and our society functioning are fatally undermined

According to hard-determinism , not only are there are no grounds for punishing bad behaviour; there are no grounds for rewarding good either. The Oscar-winning movie *Hacksaw Ridge* tells the story of the Christian and pacifist Desmond Doss, who served in World War II despite vowing never to hold a gun. After a particularly brutal onslaught from the Japanese, his platoon was forced to retreat for the night, leaving many men injured on the field. Dawes spent an exhausting night rescuing them, one by one, and lowering them down the ridge by a rope passed through his bleeding hands. Roughly 75 lives were saved by this one man. This kind of sacrifice is rare. When we see it, it stirs something deep inside us. Medals are given; acts of remembrance are instituted; movies are made. But here is the problem. If there is no libertarian free will, then there is nothing to honour. Dawes could not have done otherwise. His genes, his background and his brain activity were the drivers of his actions and nothing more. Surely bravery is bravery precisely because the person *could* have done otherwise, yet they chose not to. We do not live as hard-determinists. We live as though our choices mean something. It fails the third test.

WHY FREE WILL?

If it is true that we are free, volitional beings, then for what purpose? Simply so that we can please ourselves? Or is there a purpose beyond making life more interesting or pleasurable for ourselves; beyond being able to choose between 20 different types of cereal; beyond noble pursuits to liberate people and restore human freedom?

According to the Christian worldview, we have a level of responsible choice because we are made by God, and aspects of ourselves reflect aspects of him. In fact, the God who is revealed in the Bible, and supremely through Christ, has endowed humans with the dignity of responsible choices, even though there are many factors that influence and affect us. We each have the capacity, and the moral responsibility, to choose to love or to hate, to help or to hinder, to embrace or to reject.

Choice is integral to relationships. Romantic relationships and friendships begin and continue on the basis of decisions as well as feelings. The deeper the commitment, the deeper the significance of the decision. God is deeply committed to us and offers us relationship—one that is both similar and different to those we have with people. It is my experience, and that of countless others, that entering into friendship with God is the most liberating experience of all. God's desire is that all people would come to know him.[105] And Jesus himself said, "Come to me, all you who are weary and burdened, and I will give you rest" (Matthew 11 v 28). Friendship with him is freely available to all, but the friendship begins for those who respond to his invitation. Why? Because God is interested in relating to us as people rather than as robots.

C.S. Lewis, who was an author and professor of English Literature at Oxford University, describes his journey from Atheism to Christianity in his autobiography, *Surprised by Joy*. Having discounted God for much of his life, he began to change his mind, and even describes a particular moment of "free won't".

> *The odd thing was that before God closed in on me, I was in fact offered what now appears a moment of wholly free choice. In a sense. I was going up Headington Hill on the top of a bus. Without words and (I think) almost without images, a fact about myself was somehow presented to me. I became aware that I was holding something at bay, or shutting something out ... I felt myself being there and then, given a free choice. I could open the door or keep it shut ... The choice appeared to be momentous but it was also strangely unemotional. I was moved by no desires or fears. In a sense I was not moved by anything. I chose to open, to unbuckle, to loosen the rein. I say "I chose," yet it did not really seem possible to do the opposite. On the other hand, I was aware of no motives. You could argue that I was not a free agent, but I am more inclined to think that this came nearer to being a perfectly free act than most that I have ever done.* [106]

Our greatest freedom comes when we open the door to Jesus. Free will enables relationship with God, which may explain why some choose to undermine it. The stakes surrounding free will are high. The invitation is open to all, but the choice, and the responsibility for it, is ultimately yours.

Are we hard-wired to believe?

Wherever we go in the world, there are religious practices, from Hinduism in India, to Buddhism in Japan, to Islam in the Middle-East and parts of south-east Asia. New religions also continue to emerge, such as the remarkable and bizarre cargo cults of the South Pacific islands. After the arrival of Western civilisation in the 1940s, some islanders believed that if the correct rituals were performed, then cargo—shipments of Western riches—would be delivered by the gods. They made effigies of aircraft and ships out of straw and longed for a time when they would receive a cargo of wealth. Some islanders in Vanuatu even formed a religion that worshipped Prince Philip, the Duke of Edinburgh.

Western religious expression is hugely varied today. Many prefer to seek their inner self through yoga, self-help and New Age practices using crystals and other techniques. We can still see the tail end of dull-but-polite cultural Christianity, yet there are also thousands of young people who can be found worshipping God during summer festivals, and weekly at meetings that are built around fresh expressions of church. Many of those that identify as having "no religion" will often admit to having prayed at some point during their life. Wherever we go, we find worshippers of some sort. What does this tell us about religion? Does it mean that every person is somehow hard-wired for religious belief? And if so, does this undermine the notion of the reality of God?

NATURAL RELIGION

Cognitive science of religion (CSR) is the study of what is happening in the mind and brain during religious belief and practice, and is a thriving and growing area of research. Many cognitive scientists make the case that religion can now be explained through natural processes: what they call "natural religion". They describe how people in every culture and tradition, right from early childhood, have an in-built tendency to believe in the supernatural or be superstitious. A number of core intuitions occur across cultures and especially in young children. These include belief in a God or gods, the existence of the soul, and belief in the afterlife.[107] According to CSR, humans have a natural bias towards religious belief. In the words of one author, we are "*Wired for God*".[108]

Is there a supernatural reason for this in-built hard-wiring, or has CSR taken the "magic" out of religion by explaining it away? According to atheist Daniel Dennett, natural religion has indeed broken the spell surrounding belief in God.[109] In the dark ages, the argument goes, religion was seen as mysterious and supernatural, but now cognitive science has shown it to be a product of nature and no more.

Supernatural explanations are no longer relevant and deities even less so.

Is this true? Let's begin by looking at three different attempts to explain away religion that are not mutually exclusive. First, religion results from human error. Second, religion is a product of evolution. And third, that religion stems from our genes.

1. SADLY MISTAKEN

Children are often scared of the dark. They have a tendency to imagine monsters under their bed and visits from trolls and other similarly disturbing creatures. I remember, as a child, getting up in the dark in the middle of the night to visit the bathroom. As I entered, a black creature was sitting ominously in the corner. For some reason, I presumed it to be a cat (even though we have never had a cat) and let out a scream that woke the rest of the family. My mum rushed in and turned on the light. The mysterious cat was in fact a toilet brush.

CSR researchers believe that people have what is known as a Hypersensitive Agency Detection Device, or HADD, built into the workings of their mind. This device enables us to spot patterns and pick up signals from our surroundings. In terms of evolution, this would be helpful for recognising when we are in danger from predators. We might also expect it to allow for lots of false positives, because to be mistaken about predators would be to risk death. Better to err on the safe side.

One line of thought is that since this device is hypersensitive, it makes many errors and is unreliable. In other words, humans have a natural tendency to believe in things that simply don't exist. We are prone to errors in our thinking.

We imagine things to be real that are not. We see cats in the dark. When things go "bump in the night", we wonder if there is an intruder. We make connections between events and look for patterns when in fact there are none. We dance and pray to inanimate objects to protect us, but they are as unreal as shadows in the corner that spook us in the dark.

Scientists have described this capacity to believe in different ways. Some have attributed it to errors in the human mind.[110] Others have argued that religiosity is dictated by dopamine levels, a chemical naturally present in the brain which can lead to errors in decision-making.[111] Still others have described religious belief as a psychiatric illness. Richard Dawkins believes that theists are so badly mistaken that they are deluded,[112] so much so that he paraphrased American writer and philosopher Robert Pirsig in saying that…

> *when one person suffers from a delusion, it is called insanity. When many people suffer from a delusion it is called a Religion.*[113]

Mistaken about being mistaken?

The problem with these views is that they assume the person describing their position is not subject to the same errors. The belief that religion results from error could itself be an error. The belief that religion is a delusion could itself be a delusion. Perhaps the view that the religious are mistaken is itself making a connection that doesn't exist. They are caught in the same logical trap as the hard determinists—it becomes impossible to say anything meaningful.

Dr Michael Murray of the John Templeton Foundation makes the point that the context of a belief impacts whether or not the belief is reliable.[114] My hearing is generally reliable in helping me form beliefs about my surroundings, except when I am underwater or my ears are blocked. Context plays into reliability. In optimal conditions, we can trust our HADD to be reliable. Very often, we think we have heard a knock at the door and a person is waiting there. The sound in the night is the baby crying. Our instincts are correct. Only in sub-optimal conditions would it make sense to assume that our sensory detection antennae are faulty.

All Religions and none

A survey by the Pew Research Centre in 2015 showed that, worldwide, there are five major categories of religious belief: Christianity (31.2%), Islam (24.1%), no religious belief (which includes atheism and agnosticism)* (16%), Hinduism (15.1%) and Buddhism (0.5%).** Folk religions (0.4%) and other religions (0.1%) are myriad and bring the total number of religions into the thousands. If it is true that we possess this HADD and that it is responsible for generating religious beliefs, why would it lead to such a wide range of, often contradictory, religious possibilities? Why not just one religion? Surely this scattergun approach confirms its unreliability?

Here, we must ask an important question. Can CSR explain religious belief *entirely?* The question we asked at the beginning of this book is important again here: "Am I just my brain?" If yes, then natural explanations of religion are all that there is and will ever be. But if I am *more* than my brain, other factors will also contribute to religious belief. CSR seeks to explain why we have religious tendencies *in*

---

* Referred to as "Unaffiliated" in the Pew Research Survey, 2015.
** pewforum.org/2017/04/05/the-changing-global-religious-landscape (accessed 8.1.2019).

general, but upbringing and culture will influence the *specific* religion that we adopt. Murray comments:

> *HADD tells me there is "an agent"; my beliefs about what sorts of fauna inhabit these parts lead me to conclude that the agent is a bear or a tiger or the bogeyman. If you conclude that it is a bear and I conclude that it is the bogeyman, this doesn't show HADD to be unreliable, it shows that my mom was wrong to teach me that there is a bogeyman.*[115]

The statistics on world religions remind us that whatever the country or culture, there are worshippers to be found. In this sense, HADD is reliable.

2. SURVIVAL OF THE RELIGIOUS?

A second line of thought as to why religion might be "natural" is because it has helped the fittest to survive. According to some scientists, answers are found not in the heavens, but in our ancestors. Belief in a higher being has helped the human race survive and has been beneficial to society. How might this have happened?

First, religious belief provides cohesion in communities, especially when they are not genetically related. With a common cause, unrelated families work better together, making survival of communities more likely. Second, belief in a higher being gives a sense of meaning, purpose and well-being which is more likely to lead to reproduction.[116] Third, the perception that we are being watched enforces good behaviour, much as speed cameras cause drivers to slow down. Good behaviour leads to relational harmony and increased survival chances. Perhaps we could call this the "Big Brother is watching you" effect.[117] So far, so plausible.

But how exactly could the processes of nature yield religious beliefs? There are two main theories. The first is that religious belief is a biological *adaptation*. In other words, religion has been *directly* selected for during the course of evolution because of its benefits.[118] The second view is that religious belief is an *indirect* product of evolution.[119] The human brain has developed the ability to reason, think in abstract terms and problem-solve on behalf of communities. Religious belief has "piggy backed" upon existing processes. According to this view, religious belief is a *by-product* of evolution. Either way, scientists agree that the cognitive architecture that underpins religious belief has helped humanity survive.

Are you thinking clearly?

These theories certainly do not rule out God, and could be describing the means by which the spiritual antennae of humanity have developed. However, it is worth asking the question of why we have reliable cognitive faculties in the first place. Philosopher Alvin Plantinga has pointed out that the very existence of a trustworthy mind makes more sense if one supreme being has overseen the creative process.[120] If matter is all that exists, then the evolutionary process is "under no obligation" to generate minds capable of truthful and clear thinking about anything, let alone anything religious. The goal would be to create minds capable of survival, not cognitive reliability. Attempts to explain our cognitive faculties from within a universe of meaningless matter undermine the very minds through which those arguments came. In the words of Professor John Lennox, such an approach "shoots itself not only in the foot but also in the brain".[121]

Some counter this objection by saying that truth would be necessary for survival.[122] The truth about whether there is a nearby predator or deadly insect could be a matter of life or death. Therefore, evolution could conceivably select for accurate perception about one's surroundings in some circumstances.

This may be true, but, as C.S. Lewis and others have pointed out, materialism does not offer the *best* explanation for the rationality and higher-level thinking that humans uniquely possess. Why would a non-rational process necessarily give rise to a rational, reliable mind? Yet, if a conscious, rational being underpins the cosmos, then we have grounds for making sense of cognition. Our cognitive faculties can be reliable because God exists.

3. IS THERE A GOD GENE?

Some take the arguments for natural religion into the realm of genetics. If we are naturally biased towards religion, then has this come from our genes? In other words, is there a God gene? In 2004, geneticist Dean Hamer published a book entitled, *The God Gene: How Faith is Hardwired into Our Genes*. This idea went into wide circulation and was even picked up by *Time* magazine the same year. However, Hamer's book was shown to be lacking in areas and was viewed with scepticism in the scientific community. *Time* magazine did not report that Hamer's book had been poorly received in academia, because it doesn't make for sensational reading. But the idea that religiosity rests on a single gene is simply not true. There is no God gene.

How do we know this? We know it from multiple studies on identical twins, who have exactly the same genes but often different levels of religiosity.[123] These studies have shown that there are some mild genetic links to behaviours such as "self-forgetfulness" and "self-transcendence". However, these links are spread across many different genes rather than being tied to a single gene. Furthermore, the ability to forget oneself or think above and beyond oneself is not restricted to religious people. The irreligious are also capable of these things. The science shows that spirituality is highly complex, and therefore there is no God gene.

Interestingly, the twin studies showed that upbringing has a stronger effect than genetics. The beliefs you adopt as an adult are often the beliefs you grew up with. But even this doesn't explain how some beliefs arise. How do you explain Christian parents with non-believing children? Or non-believing parents with children who become Christians, which is the story of countless people around the world? Moreover, how do we explain the fact that some of the fastest-growing Christian communities are in countries where religion was prohibited in the previous generations? In China, Chairman Mao set out to remove religion entirely, burning religious books in library after library. Yet now, and despite continuing persecution, China has one of the fastest-growing churches in the world, estimated at around 100 million members. Upbringing plays a role but does not guarantee outcomes.

The fact that there is no God gene can be seen as both good news and bad news. The good news is that your genes are not going to force you to believe something you don't want to. The bad news is that we cannot blame our genes for our choices. Whether or not we believe in God remains a choice. Everyone has all the DNA they need to believe.

PERHAPS GOD EXISTS?

Has cognitive science of religion "broken the spell" surrounding belief in God? Not really. CSR as a scientific discipline does not ask us to draw this conclusion. Our beliefs and opinions lead us there. Natural explanations of religious tendencies say nothing about whether or not God actually exists. That would be a bit like saying knowledge of the software design behind Facebook rules out the existence of Mark Zuckerberg. It does not necessarily follow that since we understand how the mind is working, God does not exist.

Perhaps we are wired for God because God exists? Justin Barrett, Professor of Psychology at Fuller Theological Seminary, has compared our cognitive faculties to a radio.[124] Understanding how a radio works doesn't answer the question of whether there is someone "out there" broadcasting. Nor does it answer the question of whether it is right and good to tune in. It simply tells us that there is a radio capable of picking up signals. Perhaps this wiring exists because God has made us with a natural capacity to connect to himself?

Natural Christianity

Justin Barrett believes that many aspects of "natural religion" are extendable to Christian belief. In other words, there is a "naturalness" to parts of Christianity as well. For example, according to natural religion, people have a natural intuition that the universe has been made purposefully by a superhuman being. In Christianity, this translates into the belief that the universe is made by the God of the Bible. Or, the natural intuition that there is more to a person than simply their body translates into the Christian belief that humans have a soul.[125] Some of the "naturalness" of religion can be applied to the Christian faith without problem.

· The idea that all humans have an innate and natural capacity to know and relate to God is also referred to in the Bible. The apostle Paul, in his letter to the Roman church, mentions this very concept, albeit from a different angle. He says that God does not hide himself away. The invisible God is visible in nature for everyone to see. Moreover, humans have a natural ability to ponder the invisible God based on what they see around them. Paul writes,

> ...what may be known about God is plain to [people], because God has made it plain to them. For since the creation of the world God's invisible qualities – his eternal power and divine nature – have been clearly seen, being understood from what has been made, so that people are without excuse. ROMANS 1 V 19-20

According to the Christian worldview, "God's invisible qualities" can be seen in "what has been made". Natural processes do not necessitate natural origins. These verses tell us that natural processes that can be "clearly seen [and] understood", and they speak of God's "eternal power and divine nature". Natural processes point to supernatural origins, and in such a way that is plain to see.

Quid Pro Quo?

A few years ago, one of my friends had a baby, and I decided to take the family some food. With young children myself, I had very little time or energy to come up with exciting options. Eventually, I managed to pop in with some frozen pizzas and a tub of ice cream. Within a couple of weeks, this same friend, despite having a newborn, was on my doorstep with a gift for me. This exchange reminded me of something deeply ingrained in our humanity. We live in a world of give and take.

If someone extends a dinner invitation, we make a mental note to invite them back in the next few months. Friends that remember your birthday will receive a card in return. Those that don't will have a card politely withheld, even though you did remember. The people that "like" your social media posts will receive the same in kind. We seem to be wired for *quid pro quo*.

Yet Christianity centres around someone who takes the "hit" so that other people may benefit. One who loves *in spite of*, not *because of*, the love extended to him. One who bears the burden of our mistakes and regrets, despite having none himself. One who dies so that others may live, both in this life and the next. Is this act of self-less love not completely at odds with natural religion?

Evolutionary biologists claim that there are also instances of altruism in nature, said to occur when "[an organism's] behaviour benefits other organisms, at a cost to itself".[126] For example, Veret monkeys warn their troop about approaching predators by releasing a cry that helps other monkeys but increases their own chances of being hunted. Ant colonies contain sterile workers who solely tend the queen, to maximise her offspring.

Though these rare occurrences are intriguing, critics respond that these are not instances of altruism in the sense that we are meaning here[127]. Biological "cost" is measured in terms of reproductive fitness. If an organism reduces the number of their own offspring to enable another's to increase, then this is deemed to be altruistic. But this is quite different to the conscious intention to help another, to the detriment of oneself. Arguably, the animal kingdom is not capable of this kind of altruism at all. The altruism in nature that biologists speak of is very different to the selfless and sacrificial love of one person for another.

Wired for Grace?

At the heart of the Christian faith is something very unnatural. Jesus Christ has reconciled us to God by taking away the sin that separates us from him. On the cross, Jesus willingly gave his life so that the broken relationship between God and people could be mended. If it is true that God has done this for us, then our role is to decide whether or not to accept the hand of friendship.

This is known as the grace of God. But grace does not sit well in a world of *quid pro quo* and goes against the grain of natural religion.[128] Even those who have been Christians for a long time have to resist the idea that friendship with God stands or falls on the number of hours spent in prayer, reading the Bible or helping others. It stands or falls on what Jesus has done for us on the cross and whether or not we choose to accept and follow him. Our natural instinct is to climb our way to heaven through *quid pro quo* with God, but in Jesus, heaven has climbed its way down to us. The grace of God is the very thing that frees us from ourselves and frees us to be ourselves. It frees us to forgive, be forgiven and start again. Those that grasp grace can say along with the former brute and slave-trader-turned-priest John Newton:

> *Amazing Grace how sweet the sound*
> *that saved a wretch like me.*[129]

There is nothing natural about grace. Grace is more than simply "unnatural". Grace is supernatural. The same could be said for a host of other beliefs that are central to Christianity, including the virgin birth, the incarnation and the death of Jesus. Perhaps the most unnatural of all is the resurrection of Jesus, which we will say more about in chapter 7. The natural way of things is that dead people stay dead. Yet Christianity is supernatural at its core, because it insists that dead people rise—and therefore, natural religion cannot be the whole story.

Are we "hard-wired" to believe? In some ways yes, and in others no. We all have the brain equipment needed to connect to God, but neither our genes nor our cognitive faculties are forcing us to do so. We are certainly not hard-wired for the unmerited, undeserved love of God that asks only for a response. Grace is received, not earned. Grace is unnatural. Grace is supernatural.

Is religious experience just brain activity?

Temporal lobe epilepsy (TLE) is a disorder in which part of the brain is overactive, leading to seizures. Just before an episode, the person experiences what is known as an "aura", in which feelings of awe, ecstasy, and sometimes a "divine presence" can arise. Novelist Fyodor Dostoevsky was one such sufferer, and wrote about his epilepsy through novels such as *The Idiot* and *Demons*.

Neurosurgeon Wilder Penfield, whom we encountered in earlier chapters, developed a treatment for TLE known as the Montreal procedure, which involved surgical stimulation of the temporal lobe. Penfield discovered that patients sometimes described feeling "out of their body" during treatment,[130] raising some interesting questions. Was the Montreal procedure inducing a kind of religious experience?

The subject of religious experience is inaccessible to many of us. It may only cross our radar when a second cousin goes travelling to "find themselves", or when a college friend smokes something regrettable or takes up yoga. Do these observations from epilepsy patients confirm what many have believed to be true for years? That is, that religious encounters can be explained simply by the activity of the brain?

So far this century, neuroscientists have published a great deal of research showing numerous brain regions to be active during different religious practices.[131] Have these discoveries reduced religious experience to an entirely natural phenomenon? Are we just our brains once again? Is what we used to think of as an encounter with "the divine" now merely a series of brain manipulations? Has neuroscience filled a gap and squeezed out God?

Whereas the previous chapter looked at whether religious *belief* can be explained by nature, this chapter will ask similar questions of religious *experience*. Religious experience is the junction at which neuroscience filters into theology, and since the 1990s this field has been referred to as neurotheology.

WHAT DO WE MEAN BY RELIGIOUS EXPERIENCE?

Religious experience is a very broad term and encompasses a range of beliefs and practices. The classic work of psychologist William James' (1842-1910), *The Varieties of Religious Experience,*[132] is still considered one of the most important contributions to the field. According to the *Oxford Dictionary of Philosophy*, a religious experience is defined as "Any experience carrying as its content the presence of something divine or transcendent".[133] Religious experiences can be public and encountered alongside other people, or they can be private and experienced alone. They can involve features that are normal to life, such as words we understand, or they can go beyond our everyday reality, such as more mystical experiences. They can be short-lived and dramatic, or they can constitute a long-term general sense of divine guidance or peacefulness. They can occur during religious practices initiated by the person, such as prayer, meditation or singing, or they can happen without being sought out, such as the apostle Paul's Damascus-road experience (see Acts chapter 9).

Religious experience can mean many different things, and so we will focus our discussion on some particular questions.

God in the temporal lobe?

The first question we will consider is this: is there a place in the human brain assigned to religious activities and to God? Penfield had some interesting results from stimulation of the temporal lobe. Could this be a "God spot" that is active in a religious person but less so in agnostics, atheists and those of no religion? The notion of a God spot or module gained some traction in the 1990s[134] leading to books such as *Where God Lives in the Human Brain*.[135]

Many were sceptical about this idea, because it transpired that most patients did not describe the pre-seizure aura in religious terms but tended to use everyday language involving tastes, smells, memories, emotions, joy and so on. If the patients themselves did not think they were having a spiritual encounter, then notions of a "God spot" are weakened. The extent to which the aura was described as religious depended more on the pre-existing beliefs of the person rather than their brain activity *per se*. Moreover, neuroscientists today speak more of brain networks than individual "spots". Religious activity, as we will see later, engages a network of regions rather than a single, isolated lobe.

So, religious experience cannot be reduced to brain activity in the temporal lobes. Just as there is no "God gene" there is also no "God spot".

The "God Helmet"

Nevertheless, neuroscientist and philosopher Michael Persinger continued to pursue the idea of a "God spot", this time in volunteers without epilepsy. Persinger wanted to see if he could induce religious experiences in people by artificially stimulating their temporal lobes using transcranial magnetic stimulation (TMS). The signal was applied to the brain using a helmet, initially named the Koren helmet after its inventor Stanley Koren, but eventually renamed in the press as the "God helmet"—that is, the helmet that generates a "sense of God".

According to Persinger, 80% of people who wore the helmet reported a "sense of presence", which some volunteers described as mystical or religious. Furthermore, atheists who took part in the experiment gave very different results. Susan Blackmore experienced very powerful emotions, including anger, but not a "sense of presence". Richard Dawkins felt nothing except his left leg twitching.[136]

Persinger's results received mixed reviews. Some interpreted them as evidence that belief in God and religious experience were strongly connected to activity in the right temporal lobe. Others viewed the results as controversial for several reasons. First, *some* volunteers sensed "a presence" even without the TMS, and so the power of suggestion may have played a role. Second, the data were hard to replicate. Third, the TMS field strength was significantly weaker than those normally used in research and not considered strong enough to cause changes in the brain.

Even if Persinger had managed to induce an artificial

religious experience, this doesn't undermine *all* religious experiences. Philosophers argue that, if they were able to artificially stimulate the optic nerve to generate a visual experience of an apple, this would not undermine the reality of apples.[137] Rather, it would merely help us understand some of the brain regions and mechanisms involved in visual experience. In no way would we conclude that apples are illusory. The same applies to religious experience. Even if a religious experience *could* be generated artificially, it does not mean all such experiences are illusory.

Christian Prayer and Buddhist Meditation

What happens in the brain when people pray? Over the last 30 years, brain-imaging techniques have been used to look inside the brain during a host of different religious practices. Neuroscientists such as Professor Andrew Newberg have pioneered research into Buddhist meditation, rituals, trance states, Christian prayer, and also a more unusual type of Christian prayer known as "praying in tongues": in other words, praying in what is believed to be a supernatural language.[138] A review in 2009 lists 40 different brain regions involved in prayer and meditation,[139] showing that the brain is very active during every spiritual endeavour.

If the brain seems to track with all kinds of religious practice does this mean that those religions are essentially the same? Is this a case of our neurons neutralising the differences between religions? Well, the neuroscience suggests not. Studies so far indicate different patterns of activity for different types of prayer. For example, when praying to a personal being, areas associated with relationality are recruited, whereas in Buddhist meditation, which is more open-ended, different brain networks are used. Our understanding of this area is still developing, and it may be too early to draw any firm conclusions, but certainly the brain does not process all religious practices in a one-size-fits-all manner.

Questions of the validity of the different world religions reach beyond neuroscience into philosophy and theology. The questions posed in chapter 1 are a helpful framework for assessing a religion: Is it internally coherent? Does it have explanatory power? Can it be lived? The remainder of this chapter will focus specifically on religious experience in the Christian faith.

DOES BRAIN ACTIVITY MEAN THE EXPERIENCE ISN'T REAL?

The discoveries of Newberg and others raise an interesting question. Does the presence of brain activity during prayer mean it must all be in your head? Does brain activity mean the encounter isn't real?

Not necessarily.

There is far more to prayer than just brain activity. Just because we know where in the brain something might be happening, that doesn't mean we have fully explained this practice.

Lots of people love chocolate. I love chocolate. It is not just the taste that is great, but also the anticipation of the taste as we get ready to indulge. Neuroscientists now know that right from the moment you decide to eat some chocolate, a network of "pleasure" or "reward" centres starts firing and release brain chemicals that lead to the inevitable "happy place". These networks are also the same ones that go into overdrive when we are in love. They are also the same networks that spiral out of control when a person is addicted to drugs.

It is one thing to understand the brain's involvement in

chocolate consumption, but quite another to *experience* the taste of chocolate. Knowledge of the first does not rule out the existence of the second. The term "hard problem of consciousness" (discussed in chapters 3 and 4) was coined precisely because brain processes and human experience are two very different things. Brain activity does not provide a carte-blanche for eliminating genuine experience.

Furthermore, brain data tell us nothing about why the person has chosen to consume chocolate, what events in the day have led to this point, and what emotions they associate with eating in general. Perhaps someone's birthday is being celebrated. Perhaps they haven't eaten all day and are reaching for a quick-fix. Perhaps there are feelings of guilt. Equally, we might understand some of the functioning of the brain during romantic love, but that tells us nothing about whom we choose to love, and for how long or how to treat that partner each day. Even if we understand the neurobiology of drug addiction, that can't tell us how someone reaches the point where destructive chemicals have become the central focus of their life.

To determine if an encounter is authentic, we need to ask some more questions. What type of encounter is it? Is it consistent with the beliefs of the person? Are there other instances of this encounter? Can it be verified? The story of the person, and perhaps of other observers too, will be as important as the signal from their brain in deciding whether the encounter is a genuine one. The benefits of brain-imaging are huge, but there are also limits to what it can tell us. Brain activity does not mean the encounter is inauthentic.

DOES BRAIN ACTIVITY MEAN THAT GOD ISN'T REAL?

Does the presence of neuronal activity call into question the existence of God? Should Christians be fearful that neuroscience is squeezing God out of the picture? Not at all. Just because something is *experienced through* the brain, that does not necessarily mean it *originated in* the brain. The fact that we know and understand reward circuitry in the brain does not mean that we call into question the existence of chocolate. That's an absurd idea! Nor would we call into question the existence of our boyfriend, girlfriend or partner, whose love also activates our brain. The very fact that chocolate and our partner exist is why there is brain activity in the first place.

Similarly, brain activity during prayer does not negate God. In fact, philosophers such as Alston, Plantinga and Swinburne argue that authentic religious experiences more generally are evidence *for* God.[140] And if God *does* exist, then it comes as no surprise that he would make us such that our brains are active when we encounter him. This kind of data is not a threat to religious, and specifically Christian, belief.

Of course, if you believe that we are our brains, then one could argue that brain activity explains away religious experience. But if there is a distinct mind that influences the brain, then brain activity is only one of several different types of explanation. To conclude that we must choose between activity in the brain and the reality of God is to sell ourselves short. Both are needed to fully understand religious experience.

Great Expectations

For a person's brain and mind to be both engaged when they pray is exactly what we would expect. The God revealed in the pages of the Bible has made people with both physical and spiritual dimensions. A person is a mysterious and beautiful embodiment of both the physical and the spiritual working together. If encounters with God are real, we would expect them to engage the brain rather than by-pass it. Brain activity, far from being a threat to God, is exactly what we would predict. Something concrete is going on, with physical and spiritual processes involved. Theologian N.T. Wright puts it like this,

> ...God's sphere and our sphere ... are not thought of as detached or separate. They overlap and interlock. God is always at work in the world, and God is always at work in, and addressing, human beings, not only through one faculty such as the soul or spirit but through every fibre of our beings, not least our bodies. That is why I am not afraid that one day the neuroscientists might come up with a complete account of exactly which neurons fire under which circumstances, including that might indicate the person as responding to God and his love in worship, prayer and adoration. Why should the creator not relate to his creation in a thousand different ways? Why should brain, heart and body not all be wonderfully interrelated in so many ways that we need the rich language of mind, soul and spirit to begin to do justice to it all? [141]

The Bible does not refer to people as ghosts nor as brainwashed machines but as integrated physical and spiritual beings. If there were no brain activity during prayer, this would give more cause for concern!

But I don't have a religious brain!

Are some people more likely to find God than others? Is the extent to which I believe in God related to the wiring of my brain? A positive answer to this question might appeal to some because it takes the decision out of our hands and into our heads! I just don't have the right kind of brain to be a religious person. Yet, the brain-imaging data so far doesn't allow this conclusion.

In the middle of our family room at home sits a table. By name it is a dining table, yet in reality it serves many functions. Yes, we eat meals at it with family and friends, but the children also complete homework and various craft activities on it. We have held meetings around this table, and I have even written some of the words of this book on it. The table does not have one sole function; it can serve many different functions. Depending on the time of day, it is an office, a meeting place, a feeding station or a space for the creative arts.

The same is true of the brain regions employed during prayer; none of them are unique to spiritual activities. All serve multiple roles in the brain but are recruited during religious practice as well. Are some people more able to engage with God than others in terms of the makeup of their brain? No. Every person has the machinery they need.

WHY DOES IT EVEN MATTER?

Many of us would say that we are simply not religious people, and have no desire to be. Even if we were in search of a mountain-top experience, Christianity is the last place we would look. The Western church is often known more for its arid services and declining numbers than for its appeal to those who might be seeking God. What does the notion of religious encounter have to say to those who identify as having no religion?

The Christian faith speaks of something much more extraordinary, and in a sense also more ordinary, than this. Christianity does not stand or fall on religious experiences, important though they are. Christianity is anchored in human history and pivots around the life, death and resurrection of Jesus Christ. Did Jesus, the God-man rise from the dead, never to die again? If this happened, then it changes everything. If it didn't, then Christians are sadly mistaken. Of course, many creative ideas have been offered as to what might have happened to Jesus' body. This is not the place to unpack all the alternatives, except to say that many sceptics, including a lawyer who set out to disprove the resurrection, ended up concluding that Jesus *must* have risen from the dead.[142] If you have not yet given the evidence for the resurrection of Jesus a fair hearing, then consider doing so.

My husband, Conrad, has not always been a Christian. Conrad changed his mind at the age of 19, somewhat reluctantly. Why? He came to realise that *the resurrection of Jesus was true.* You might ask, how could a thinking person come to such a verdict? It was because of the dramatic change in Jesus' followers after seeing the risen Jesus. On the day that Jesus died they hid, scared, defensive and in fear of their lives. Six weeks later, they were willing to risk everything to get the message out that Jesus had risen. They fearlessly spoke in the open air and before authorities, regardless of the risk of imprisonment, torture and execution. One way or another, the most prominent figures in early Christian life were prepared to die for their belief. What would it take to bring about such a dramatic change in them? I believe that seeing your mentor, friend and Lord get up from his grave and live again would probably do it.

You may be asking, what relevance does an event 2,000 years ago have to do with 21st-century life and brain activity? It is relevant because the notion of religious experience seems inaccessible and undesirable to many of us. Yet there is an invitation to know God that is open to all. We are not invited into some undefined, formal, long-distance correspondence with God, nor a detached mystical experience. The invitation is to be raised from the dead. And this is not just a future hope, but a present experienced reality. To be a Christian is to turn away from everything wrong and invite the same Holy Spirit that raised Christ from the dead into our lives. The apostle Paul, in a letter to the church in Rome, put it like this:

> *And if the Spirit of him who raised Jesus from the dead is living in you, he who raised Christ from the dead will also give life to your mortal bodies because of his Spirit who lives in you.* ROMANS 8 V 11

The Holy Spirit begins to bring dead things to life in us. He gives "life to our mortal bodies". What might this look like? A dead-end job or a difficult boss might become more bearable. Maybe we begin to reconnect with our estranged family. Perhaps we are freed from an addiction or destructive behaviour. There is strength to cope with the pressure of work and exams. God is encountered in very real ways as he transforms ordinary situations and ordinary people, through his Holy Spirit. Even our own attitudes begin to change. *We* change, and for the better. All Christians experience God through the Holy Spirit who lives in them. This can resemble anything from a deep sense of peace in place of panic to a dramatic and overwhelming experience of the presence of God.

One day, the full implications of this transformation will be realised. One day God will make everything new. One day, those that know God will continue to know Him forever. One day, God will be seen and encountered face-to-face.

THE INVITATION IS FOR ALL

I will never forget the day when a disabled boy was baptised in my local church. I do not know the exact nature of his disability, except to say that he needed a wheelchair and was able to speak only through voice-recognition software. It was incredibly moving to hear him prepare for baptism by responding to the questions: "Do you turn to Christ?" "Do you renounce evil?" "Do you repent of your sins?" After each question, he answered and in a manner that clearly showed he fully understood what was happening and why. This baptism was a reminder to me that relationship with God is not dependent on having a fully functioning brain. People with impaired brain-function can still know and be known by God.

A young man, Luke, who is mentally and physically disabled, has been part of our church since birth. Sometimes during the service Luke is still, and sometimes he makes rhythmic noises while rocking backwards and forwards. Occasionally, particularly during the sermon, prayers or a particular song, he claps. For some, this may simply seem to be a case of a disabled person being stimulated by what they have heard or seen. At heart, I believe that Luke *gets it*. His parents, and those in the church family who have helped Luke over his 19 years, speak of how he is constantly open to God. Countless people have approached Luke at the end of a service to thank him for opening their eyes to new things. If you were to look into Luke's eyes, you would see God's love shining through. He knows God; he experiences him despite being mentally and physically disabled, and he expresses it without inhibition. Who knows which brain networks are firing at that moment? But one thing is clear anyone, whatever their impairment, can still know God. The able-bodied and able-minded have much to learn from people with disabilities. God is greater than the human brain, and does relate to anyone and everyone, regardless of their cognitive capacity. No one is beyond his reach.

Why can 8 I think?

Let's finish this book where it began, with a child sitting by a window on a rainy day, watching the drops splash against the pane. The child, if you remember, was me. In my slightly bored state, some questions popped into my head. "Why can I think?" "Why do I exist?" "Why am I a living, breathing, conscious person who experiences life?" It seemed that the questions came from nowhere. Unprompted. Maybe, even unwanted. I wasn't searching. I was just sitting. Yet, there the questions were. *Why can I think?* Perhaps you have experienced something similar as well.

So far, this book has attempted to look at some of the different questions relating to human consciousness: *Am I just my brain? Is belief in the soul out-of-date? Are we just machines? Are we more than machines? Is free will just an illusion? Are we hard-wired to believe? Is religious experience just brain activity?* Some have argued that because we live in a world of meaningless matter, then the answer to all these questions must be "yes". I have argued that if there is more to this world than we can see with our eyes, then the answer to each of these questions is a firm "no". Human beings are so much more than simply their brains. Brains don't think: *people* think using their brains. Humans possess a conscious mind that interacts with the brain but is not identical to it. Even my childhood intuition points us in this direction. I was the one doing the thinking, not my brain!

The discoveries of neuroscience do not confine us to the conclusion that a person *is* their brain. There are other views held by scientists and philosophers that I think make more sense of the human person and our experience of ourselves. These have been outlined in chapter 4. The discoveries of neuroscience are also entirely compatible with the existence of God, and in no way does belief in one rule out the other. Neuroscience describes the processes going on in the brain when we think, but it cannot answer the question that I had sitting by the window: "*Why* can I think?" There are questions that neuroscience cannot answer and was never intended to answer.

WHAT'S THE POINT OF CONSCIOUSNESS?

A great deal of time has been taken up in trying to answer the question of what consciousness actually *is*. But, if most of us can at least agree that it does exist, then perhaps a question to break the philosophical stalemate is to ask, what is consciousness *for*? *Why* does consciousness exist? In other words, *why* exactly can I think?

A recent edition of *New Scientist* included an article that posed this very question. They asked, "Why be conscious: The improbable origins of our unique mind". This journal is known for its atheistic stance, yet a few pages in the article admitted the limits of our current understanding and the need to ask some different questions about consciousness. According to author Bob Holmes, to move forwards, we need to ask questions of origins and purpose:

> *There is no distinctive pattern of brain activity that indicates consciousness ... We don't even fully understand what consciousness is. But maybe there's a way to get a handle on it. What if we tracked consciousness to its origins? Then, instead of asking what consciousness is, we ask why it evolved—in other words, what is it for?* [143]

Can we trace consciousness back to its origins? The question is a good one. Of course, beliefs determine how far back we look. If we believe the natural world is all there is, then our search for the origins of consciousness will remain within this world. Unsurprisingly, the article goes on to look for traces of consciousness in the evolutionary history of the animal kingdom. But what if there is more to this world than simply animals, vegetables and minerals? What if the origins of consciousness are more ancient than this? If so, then we must expand the scope of our search beyond the natural world.

In chapter 4, we asked, what if the origins of consciousness can be traced to a conscious being known as God who has always existed? If this is true, how does it help us? It means that although in the created world consciousness, mind and brain are mysteriously merged, it is also possible to be conscious *without* a body. It is possible to have a mind without a body. Before there was anything physical, there was God. And, as Jesus said, God is Spirit (John 4 v 24). This may seem a controversial thing to say in our modern age. After all, the idea behind Madonna's declaration in

1985 that "We are living in a material world" has tightened its hold over the last three decades.[144] And yet, before we dismiss this out of hand, let's at least keep an open scientific mind and see if this helps us makes sense of a few more things.

WHAT'S SPECIAL ABOUT PEOPLE?

Regardless of our beliefs, I think most of us would agree that human life is precious. We baulk at anything that demeans it, from human-trafficking to child abuse to cyber-bullying. But why? Where does this intrinsic worth come from? According to the Bible, humans are said to be made in the image or likeness of God (Genesis 1 v 27). What does it mean to be made "in the image" of something else? In popular parlance we talk of children being "the spitting image" of one of their parents. In other words, the child-parent resemblance is very strong. In the same way, a brain scan produces a detailed picture of your brain. It is not the brain itself but a picture that closely resembles the brain. We understand that the image both references and reveals the brain inside your head.

Similarly, we are said to be made in the image of God, because, in some senses, we reflect what he is like. Take relationships, for example. We don't need to look far to see that we are relational beings. What movie or Netflix box set doesn't centre around the breaking, forming, or reforming of relationships between friends, spouses, boyfriends, girlfriends, children, parents and so on? Teenagers long to be accepted into friendship groups. And loneliness is like a cancer in our individualised Western world despite the widespread availability of social media. We are relational beings. But why? Simply for reproductive purposes? Not so.

The idea of being made in the image of God lays claim to a bigger explanation: we are relational because God is relational. God has always existed as a community of Father, Son (Jesus) and Holy Spirit: one God in three Persons.

A Hebrew understanding of "image" would also have included a level of individual and moral responsibility. Humans are more than advanced apes. In fact, one psalmist describes us as closer to angels than animals, saying:

> *You [God] have made them [humans] a little lower*
> *than the heavenly beings and crowned them with glory*
> *and honour".*   PSALM 8 V 5

Humans are God's representatives on earth, and, as such, have a role in tending the natural world. The image of God helps us answer the question, "Am I just my brain?" If we are made in the image of God, then our core identity is not subject to the vagaries of a degenerative disease or age-related atrophy. Each human is infinitely precious and loved by God, regardless of what is happening to their body or brain. Each one of us is made in God's image for a life of meaning and purpose.

The concept of the image of God also helps us to answer the question "Why can I think?" We have a mind because God has a mind. We think because he thinks. We are conscious because he is conscious. And our minds our conscious awareness of self and world—though real enough—are only the beginning.

IS THERE ANYONE OUT THERE?

Most household televisions today are digital and high definition. However, those of a certain age will remember the days of analogue TV when the aerial, rather than being built-in as it is today, sat on top of the box. The TV needed to be tuned into each of the four available channels, and the aerial was key in picking up the signal. Often, the best picture was only found when someone was holding the aerial in a far-flung corner of the room. Thankfully, our aerial-holding days are long gone; however, this concept of the aerial helps us in our understanding of consciousness.

In our material world, mind and consciousness are often discussed as though they are electronic—similar to the electrical inner workings of a TV. But what if consciousness is more analogous to the *aerial* than the electronics?[145] Perhaps consciousness is better described as the mediator of myriad signals into, out of and around the brain? Even materialists are sympathetic to this view, and see human consciousness as a kind of "central control"; only, in their view, the kinds of signals are limited to the material world. But if God exists, then there are signals coming from beyond the material world as well as from within it: signals from God himself.

What's the point of consciousness? *So that we can know God.*

Close Encounters

My friend and colleague David Bennett was an atheist and gay-rights activist in his early twenties. His book *A War of Loves* recounts the impact of meeting a film-maker, Madeline, who was unapologetic that her motivation and inspiration for film-making came from her Christian faith.[146] David's initial reaction to this was profound disagreement. He believed that God was a fun-spoiling, rule-giving slave-driver who keeps people at arm's length. The Christian faith held nothing attractive for David. Then came a key question from Madeline: "But David, have you experienced the love of God?" According to Madeline, God was not only objectively true but also experientially real.

The Christian faith is not just rooted in history, philosophy and theology. God is also a person to be encountered. He is a first-person experience, not a third-person observation. Many people draw their conclusions about God based entirely on third-person observations about him, and about those who believe in him, which may or may not be correct. But God can be *known* and offers us relationship with himself. Peoples' experience of God is often very different to what they have heard *"on the grapevine"* about him. This was certainly true of David, who, shortly after this challenge from Madeline, became a follower of Jesus.

We are made for more than this world

Where does all this talk about mind, brain and consciousness leave us? In short, the answer we give to "Am I just my brain?" affects how we live today. If the mind is entirely material, it is ultimately temporary. We must use it to its maximum now, because one day it will cease to exist. And life is short. According to Steven Pinker, this is an impetus to live well. Pinker writes:

> *Think, too, about why we sometimes remind ourselves that "life is short". It is an impetus to extend a gesture of affection to a loved one, to bury the hatchet in a pointless dispute, to use time productively rather than squander it. I would argue that nothing gives life more purpose than the realization that every moment of consciousness is a precious and fragile gift.*[147]

That's one view, and some people live this way, to their

credit. But the belief that life is short and that no one is watching could just as easily provide an impetus to live badly. Many are single-minded in their pursuit of success, regardless of who is trampled on in the process. People want to leave their mark in this world, precisely because life is short. So, the view that life is short, and that consciousness is precious but fragile, does not necessarily lead to the kind of world that Pinker describes. But, what if it is true that the origins of consciousness lie beyond this world? Perhaps humans possess consciousness because we are also made for another world?

What kind of world, you might ask? Heaven? Yet, heaven is unattractive to many because it is considered to entail a disembodied, ghostly, floaty, potentially spooky existence, which is not a particularly desirable prospect. Or else heaven is depicted as a sickly sweet place inhabited by cherubs floating on clouds. In fact, this depiction stems more from Plato's idea of the disembodied, immaterial and immortal soul that we discussed in chapter 2 than from a biblical understanding of heaven and eternity.

HEAVENLY BODIES

Quite a different picture of life beyond the grave is painted in the pages of the Bible, as theologians such as N.T. Wright have explained.[148] Life in eternity will be embodied just as life on earth is now. We will not float up to heaven, but a new heaven and new earth will come down to us (Revelation 21). All of our senses will be engaged. We will be able to see, hear, touch, speak and taste. Heaven will be every bit as real as the world we experience today. It will be bigger, better, more beautiful and more real than anything we could imagine.

Some of the final verses of the Bible say that God will make everything new. In the new heaven and new earth, God will restore a physical, material world. He will resurrect and renew our bodies, our brains and our minds.

How does the resurrection of the body marry with the different views that Christians hold on the mind? In the non-reductive physicalist view, in which the mind is contingent on the brain, God will recreate the person all over again after they die. How exactly this will happen is hard to say from our vantage point. The love of God, rather than a person's brain-state, is central to their continuation. Furthermore, a God capable of creating a universe from nothing would certainly have no difficulty re-creating each individual.

In the substance-dualist view, in which the mind is an entity that is distinct from the body, the mind of the person is held by God in a temporarily disembodied state after the body dies. A new resurrection body will later be given at the renewal of all things. Either way, Christians do not believe we float off to heaven. Humans are embodied in this life and in the next. "Heaven" is concrete and good.

Freddie Mercury asked the question in 1986, "Who wants to live for ever?". Not everyone does. Many, like Pinker, rate quality of life above length. Others apply cosmetics and consume pills to slow down the inevitable ageing process. Transhumanists seek to upgrade themselves to bionic machine-human hybrids. A select few have even had their bodies frozen in the hope that one day science and technology will be able to reverse death itself. The desire to live for ever is undeniable. Why is this? The writer of Ecclesiastes makes sense of this longing by identifying that...

He [God] has made everything beautiful in its time.
He has also set eternity in the human heart; yet no one
can fathom what God has done from beginning to end.
ECCLESIASTES 3 V 11

God has placed in us a longing for eternity. Yet the scope
of what God has done is beyond what the human mind can
conceive. There is a sense in which consciousness itself is
unfathomable, but it is still a beautiful and wonderful thing.
What's the point of it? Consciousness exists as the gateway
to all of our senses. The gateway through which we experience life on earth, and the gateway through which we can
experience and know God. Both today and for ever.

ASKING QUESTIONS

The subject of this book—"Am I just my brain?"—is an important question to ask. People have arrived at very different answers. But we all have questions. Some are front and
centre in our thinking. Others are more of a "slow burn". As
a child I found myself asking, *"Why can I think?"*. That initial question led, in time, to many other questions.

One that was simmering in the background as I arrived
at university was, "Surely you can't be a scientist and believe in God at the same time?" During my first week in
Bristol, I had the opportunity to ask this question directly,
by attending an event in my hall of residence advertised as
"Grill a Christian". It was after dinner; four Christians sat in
a row at the front of the room and allowed themselves to be
"grilled" on life, God and the universe for about two hours.
I was sitting roughly three rows from the front. Halfway
through the evening, I plucked up courage, put my hand up
and asked my question. I discovered in their thoughtful and

honest answers that, yes, you can be a rigorous scientist and believe in God. This answer was surprising, but from that moment on, belief in God for me became theoretically possible. For a year or so it remained a theory. I began to spend time with people who were Christians. I watched how they lived, how they treated people and what they said. I asked a lot more questions. I grilled quite a few more Christians.

Roughly halfway through my biochemistry degree, I reached a point where, although my questions were not exhaustively answered, Christianity made sense. It made sense of why I could think, why science was possible, why people suffer, and why I am not the perfect person I would like to be despite my best efforts. I also reached the conclusion that Jesus had risen, and I came to realise that he wanted a friendship with me. My questions led me to a person.

In the spring of 1995, I gave my life to Jesus Christ; I became a Christian. I moved from being an observer of God to being his friend. I moved from knowing about God to encountering him in the first person. This did not happen overnight. There was no dramatic experience involved, but my life began to take on a peace and sense of wholeness that I hadn't known before. I came to realise that God had given me the very mind I used as a scientist. Studying the world that God had made augmented my love for biochemistry and my sense of purpose and meaning. I also became fascinated enough to read the Bible. What I had once thought of as a dull and impenetrable book became this historical yet living text that seemed to speak to my own circumstances as I read it.

I also experienced the power of forgiveness. As I admitted my wrong thoughts and actions to God, God forgave me, and I was then able to extend forgiveness to others. I experienced a new freedom in giving my bitterness over to God. I experienced joy. I have now been walking with God for more years than I walked without him, and have never regretted that decision. In fact, I wish someone had told me earlier.

CONTINUING THE CONVERSATION

You may be someone who has all kinds of questions: some of them in the front of your mind, others rumbling along under the surface. My advice to you is this: ask them—*all* of them. There may be obvious people you can go to, but if not, there are a number of ways to keep the conversation going. Let me suggest a few.

First, there are online articles and videos available at zachariastrust.org, theocca.org and bethinking.org. For teens, rebootglobal.org takes a look at some of the most pressing questions that young people are asking.

A second option is to look up a Christianity Explored (christianityexplored.org) or Alpha (alpha.org) course in your local area. These informal and friendly meetings are a great way to ask questions along with others who are exploring the Christian faith. You may have received an invitation to one in the past. Perhaps it's time to accept.

Third, you might be ready to take a look for yourself at one of the four biographies of Jesus, written by Matthew, Mark, Luke and John. Perhaps someone has given you a Bible in the past. Why not pick it up and start to read, or call into your local church to get a copy?

Fourth, it might be that, like me, you have reached the point at which it makes more sense to begin walking with God than to continue walking without him. There is no fixed formula for how to do this, except to begin to talk to God.

If you are just your brain, then you are made only for this world, and the only mantra to live by is to live well and make the most of life while you have it. Christianity says you are more than your brain—you are made for eternity. One way or another, there will be consciousness in eternity, either with Christ or apart from him. Live today with eternity in mind. God does not segment people into mind, soul and brain. He is in the business of putting people back together again. He makes people whole. Will you choose to live life with him? Whatever you happen to believe about the mind and brain, you will not regret it.

FURTHER READING

- Susan Blackmore, *Conversations on Consciousness* (OUP, 2005). A series of conversations between some of the world's leading philosophers and scientists, providing a helpful summary of non-theistic approaches to consciousness.
- Joel Green and Stuart Palmer, *In Search of the Soul: Four Views of the Mind-body Problem* (IVP Academic, 2005). A helpful book that presents and critiques four views of the mind-body problem and references many other helpful books.
- Sam Harris, *Free Will* (Free Press, 2012). A short but hard-hitting book in which Harris outlines his radical views on free will.
- Thomas Nagel, *Mind and Cosmos: Why the Materialist Neo-Darwinian Conception of Nature Is Almost Certainly False* (OUP, 2012). A fascinating read from a materialist who argues that we need to look for new ways of thinking about the mind-body problem, beyond simply material things.
- Adrian Owen, *Into the Grey Zone: A Neuroscientists Explores the Border Between Life and Death* (Faber & Faber, 2017). A neuroscientist recounts his studies of patients in a vegetative state and the astonishing discovery that some appear to be conscious within a deeply damaged brain.
- J.P. Moreland and Scott Rae, *Body and Soul: Human Nature and the Crisis in Ethics* (IVP, 2000). A careful and thorough treatment of the case for body-soul dualism from a Christian perspective.
- Malcolm Jeeves and Warren S. Brown, *Neuroscience, Psychology and Religion: Illusions, Delusions and Realities*

About Human Nature (Templeton Press, 2009). A discussion of current views about mind and brain, interwoven with Christian perspectives on human nature.

- Jonathan J. Loose, Angus J. L. Menuge and J. P. Moreland (editors), *The Blackwell Companion to Substance Dualism* (Wiley-Blackwell, 2018). A collection of essays from leading scholars in philosophy of mind. This book presents a case for substance dualism, but also critiques it and contrasts it with physicalist alternatives.

- R.J. Berry, *Real Scientists, Real Faith* (Lion, 2009). Stories of 18 scientists who have an active Christian faith.

- *The Bible.* If you have never read the Bible before, then it's well worth taking a look for yourself, starting with one of the biographies of Jesus: Matthew, Mark, Luke or John.

- Lee Strobel, *The Case for Christ: A Journalist's Personal Investigation of the Evidence for Jesus* (Zondervan, 2016). Strobel started his investigation as an atheist interviewing dozens of experts from all disciplines and walks of life.

BY THE SAME AUTHOR

- Sharon Dirckx, *Why?: Looking at God, Evil and Personal Suffering* (IVP, 2013). In this award-winning book Sharon looks at the difficult questions people ask about suffering, and interweaves them with stories of people who believe in God who have also suffered greatly.

ENDNOTES

1   M. Robinson, *Absence of Mind* (Yale University Press, 2010), p 32.

2   www.bebrainfit.com (accessed 21.1.2019).

3   R. Tallis, *Aping Mankind: Neuromania, Darwinitis and the Misrepresentation of Humanity* (Acumen, 2011).

4   Hippocrates of Cos, *The Sacred Disease* (Loeb Classics Library 148), p 174-175.

5   C. Blakemore, *The Mind Machine* (BBC Books, 1990), p 270.

6   Oxford Dictionaries define it as "The element of a person that enables them to be aware of the world and their experiences, to think, and to feel; the faculty of consciousness and thought." oxforddictionaries.com. (accessed 21.1.2019).

7   J. Heil, *Philosophy of Mind: A Contemporary Introduction* (Routledge, 2013), p 183-184.

8   For a fuller discussion of this see John C. Lennox, *Can Science Explain Everything?* (The Good Book Company, 2019)

9   Ravi Zacharias, *The Real Face of Atheism* (Baker Books, 1990), p173-178. Nick Pollard, *Evangelism Made Slightly Less Difficult* (IVP, 1997), p 47-70 (also available at bethinking.org/apologetics/deconstructing-a-worldview).

10   Stewart Goetz and Charles Taliaferro, *A Brief History of the Soul* (Wiley-Blackwell, 2011), p 26.

11   "Weighing Human Souls—The 21 Grams Theory", www.historicmysteries.com/the-21-gram-soul-theory (accessed 8th January 2019).

12   From the previous source.

13   "A Soul's Weight", *New York Times* archived article, March 11 1907, from www.lostmag.com/issue1/soulsweight.php (accessed 8th January 2019).

14   Ian Sample, "Is there lightness after death?" *The Guardian*, 19th February 2004.

15   D.C. Dennett, *Freedom Evolves* (Viking, 2003), p 1.

16   Stephen Pinker, *How the Mind Works*, (Penguin Books, 1999), p 64.

17   S. Goetz and C. Taliaferro, *A Brief History of the Soul* (Wiley-Blackwell, 2011), p 7.

18   J.P. Moreland and S.B. Rae, *Body and Soul: Human Nature and the Crisis in Ethics* (InterVarsity Press, 2000); Keith Ward, *In Defence of the Soul* (Oneworld, 1998); Richard Swinburne, *The Evolution of the Soul* (Clarendon Press, 2007); D. Willard, "On the Texture and Substance of the Human Soul", Biola University, 22nd November 1994.

19   Eleanor Stump, "Non-Cartesian Substance Dualism and Materialism without Reductionism", in *Faith and Philosophy*, 12:505-531, from S. Goetz and C. Taliaferro, *A Brief History of the Soul* (Wiley-Blackwell, 2011), p 55.

20   Nancy Murphy, *Whatever Happened to the Soul* (Fortress Press, 1998), p 27; Malcolm Jeeves and Warren Brown. *Neuroscience, Psychology and Religion* (Templeton Foundation Press, 2009); Malcolm Jeeves, *The Emergence of Personhood* (Eerdmans, 2015); Joel Green and Stuart Palmer, *In Search of the Soul* (IVP Academic, 2005); and W.S. Brown, Nancy Murphy and H.N. Malony, *Whatever Happened to the Soul? Scientific and Theological Portraits of Human Nature* (International Society for Science and Religion, 2007).

21   C. Jenner and B. Bissinger, *The Secrets of My Life* (Trapeze, 2017).

22   www.theguardian.com/science/2017/oct/18/its-able-to-create-knowledge-itself-google-unveils-ai-learns-all-on-its-own.

23   B. Haas, "Chinese Man 'Marries' Robot he Built Himself", *The Guardian*, 4th April 2017.

24   www.turing.org.uk/scrapbook/test.html (accessed 8th January 2019).

25   "Computer simulating 13-year-old boy becomes first to pass Turing test", The Guardian, 9th June 2014.

26   J. Searle, "Minds, Brains and Programs", in *Behavioral and Brain Sciences*, 1980, 3: p 417-57

27   Thomas Nagel, "What Is It Like to Be a Bat?", in *The Philosophical Review*, 1974, 83(4):440.

28   3rd September 2016.

29   Royal Society Fellow, CBE.

30   S. Greenfield, "The Neuroscience of Consciousness", University of Melbourne, 27th November 2012.

31   D.J. Chalmers, *The Character of Consciousness* (Oxford University Press, 2010), p xiv.

32   F. Jackson, "Epiphenomenal Qualia", *Philosophical Quarterly* 1982, 32:127-136; F. Jackson, "What Mary Didn't Know", *Journal of Philosophy*, 1986, 83:291-295.

33   G.W. Leibniz, *Philosophical Papers and Letters* (Reidel, 1969), p 308, from J.P. Moreland and S.B. Rae, *Body and Soul: Human Nature and the Crisis in Ethics* (InterVarsity Press, 2000), p 56-59.

34   www.nobelprize.org/prizes/medicine/2000/kandel/facts/ (accessed 8th January 2019).

35   However, some, such as Norman Doidge may say that this it is a natural property of the brain rather than mind. See his books *The Brain's Way of Healing* (Allen Lane, 2015) and *The Brain That Heals Itself* (Penguin, 2008). See also D. Evans, *Placebo: The Belief Effect* (HarperCollins, 2003).

36   S. O'Sullivan, *It's All in Your Head: True Stories of Imaginary Illness* (Vintage, 2016), p 8.

37   O'Sullivan, p 6.

38   O'Sullivan, p 7-8.

39   www.susanblackmore.uk/articles/there-is-no-stream-of-consciousness (accessed 8th January 2019).

40   Michael Graziano of Princeton University.

41   bigthink.com/think-again-podcast/daniel-dennett-nil-thinking-about-thinking-about-thinking-nil-think-again-podcast-91; and youtube.com/watch?v=R-Nj_rEqkyQ

42   Daniel Dennett, "Explaining the 'Magic' of Consciousness", *Journal of Cultural and Evolutionary Psychology*, 2003, 1(1):7-8.

43   www.goodreads.com/quotes/214805-there-s-nothing-i-like-less-than-bad-arguments-for-a (accessed 8th January 2019).

44   www.susanblackmore.uk/articles/there-is-no-stream-of-consciousness/ (accessed 8th January 2019).

45   www.theguardian.com/science/2015/jan/21/-sp-why-cant-worlds-greatest-minds-solve-mystery-consciousness (accessed 8th January 2019).

46   Christoph Koch: Allen Institute for Brain Science, Seattle; www.scientificamerican.com/article/what-is-consciousness/ (accessed 8th January 2019).

47   Adrian Owen, *Into the Grey Zone* (Guardian Faber Publishing, 2017), *p 1-3.*

48   A.M. Owen *et al,* "Detecting Awareness in the Vegetative State", *Science* 313, 1402-1402 (2006).

49   theguardian.com/news/2017/sep/05/how-science-found-a-way-to-help-coma-patients-communicate (accessed 8th January 2019).

50   S. Dieguez, "Cotard Syndrome" in *Frontiers of Neurology and Neuroscience,* 2018, 42:23-34.

51   J.B. Green, S. Goetz, W. Hasker, N.C. Murphy and K. Corcoran, *In Search of the Soul: Four Views of the Mind-Body Problem* (Wipf & Stock, 2010), p 22. Richard Swinburne also unpacks a similar thought experiment in his book *The Evolution of the Soul* (OUP USA, 1997), p 147-149.

52   A number of non-theists hold the NRP view, including Berkeley Professor of Philosophy, John Searle; Oxford Professor of Philosophy, Baroness Susan Greenfield; and Sir Roger Penrose, Professor of Mathematics, also at Oxford.

53   Christian proponents of NRP include Nancy Murphy, Professor of Christian Philosophy at Fuller Theological Seminary; Professor John Polkinghorne, KBE FRS, theoretical physicist, theologian, writer and Anglican priest at the University of Cambridge; Professor Malcolm Jeeves, CBE FRSE FMedSci, Emeritus Professor of Psychology at the University of St. Andrews; and Peter van Inwagen, Professor of Philosophy at the University of Notre Dame.

54   For example, H. Hudson, *A Materialist Metaphysics of the Human Person* (Cornell University Press, 2001), p172-192; and L.R. Baker, "Need a Christian Be a Mind/Body Dualist?" in *Faith and Philosophy,* 1995, 12(4):489-504.

55   I. Tattersal, "Human Evolution: Personhood and Emergence" in M.A. Jeeves and D. Tutu (editors), *The Emergence of Personhood: A Quantum Leap?* (Eerdmans, 2015), p 44.

56   K. Matsumoto, W. Suzuki and K. Tanaka, "Neuronal Correlates of Goal-based Motor Selection in the Prefrontal Cortex"' *Science,* 2003, 301(5630):229-32.

57   F. Swain, *The Universe Next Door: A Journey Through 55 Parallel Worlds and Possible Futures* (John Murray, 2017), p 166.

58   There is a discussion of this topic in R.A. Varghese (editor), *The Missing Link: A Symposium on Darwin's Framework for a Creation-evolution Solution* (University Press of America, 2013), p x-xiv.

59   J.P. Moreland, *Consciousness and the Existence of God: A Theistic Argument.* (Routledge, 2009), p 15-16.

60   Michael Sabom, *Light and Death* (Zondervan, 1998).

61   R.A. Moody, *Life After Life* (HarperOne, 2015); E. Kübler-Ross, *On Children and Death: How Children and Their Parents Can and Do Cope with Death* (Simon & Schuster, 1997); M.B. Sabom, *Recollections of Death: A Medical Investigation* (Harper & Row, 1981); P. Van Lommel, "About the Continuity of Our Consciousness" in C. Machado and D.A. Shewmon (editors), *Brain Death and Disorders of*

Consciousness (Kluwer Academic/Plenum, 2004); S. Parnia, D.G. Waller, R. Yeates and P. Fenwick, "A Qualitative and Quantitative Study of the Incidence, Features and Aetiology of Near Death Experiences in Cardiac Arrest Survivors", *Resuscitation*, 2001,48(2):149-56; B. Greyson, "Incidence and Correlates of Near-Death Experiences in a Cardiac Care Unit", *General Hospital Psychiatry* 2003, 25(4).

62 K. Ring, *Life at Death: A Scientific Investigation of the Near-Death Experience* (Coward, McCann & Geoghegan, 1980); K. Osis and H. Erlendur. *At the Hour of Death* (Avon, 1977).

63 M. Morse and P. Perry, *Closer to the Light: Learning from Children's Near-Death Experiences* (Bantam, 1992).

64 Pim van Lommel, *Consciousness Beyond Life* (HarperOne, 2011); Jan Holden, Bruce Greyson and Debbie James, *The Handbook of NDEs: Thirty Years of Investigation* (Praeger, 2009).

65 Jeffery Long and Paul Perry, *Evidence of the After-life: The Science of Near-Death Experiences* (HarperOne, 2009).

66 G.R. Habermas and J.P. Moreland, *Beyond Death: Exploring the Evidence for Immortality* (Wipf & Stock, 2004).

67 A. Griffin, "Brain Activity Appears to Continue After People are Dead, According to New Study", *The Independent*, 9th March 2017.

68 G.R. Habermas and J.P. Moreland, *Beyond Death*, p 155-172.

69 K.L. Woodward, "There is life After Death", *McCall's*, Aug 1976, p.136; J. Kerby Anderson, *Life, Death and Beyond* (Zondervan, 1980), p 91; Elisabeth Kubler-Ross, *On Children and Death* (Macmillan/Collier Books, 1983), p 208, from G.R Habermas and J.P. Moreland, *Beyond Death*, p158.

70 John Audette, "Denver Cardiologist Discloses Findings After 18 Years of Near-Death Research," *Anabiosis*, Volume 1 (1979), p 1-2; Dina Ingber, "Visions of an Afterlife," *Science Digest*, (89, 1, Jan.-Feb. 1981), p 94-97, 142, from G.R Habermas and J.P. Moreland, *Beyond Death*, p160.

71 www.express.co.uk/news/science/868086/LIFE-AFTER-DEATH-What-happens-when-you-die-near-death-experience-NDE (accessed 8th January 2019).

72 E. Alexander, *Proof of Heaven: A Neurosurgeon's Journey into the Afterlife* (Piatkus, 2013).

73 D.J. Chalmers, *The Character of Consciousness* (Oxford University Press, 2010), p 15-17.

74 W. Jaworski, *Philosophy of Mind: A Comprehensive Introduction* (Wiley-Blackwell, 2011), p 231.

75 Thomas Nagel, *Mind and Cosmos: Why the Materialist Neo-Darwinian Conception of Nature is Almost Certainly False* (Oxford University Press, 2012), p 68-69.

76 Anil Ananthaswamy, *New Scientist*, 3rd September 2016.

77 Richard Swinburne, *Mind, Brain, and Free Will* (Oxford University Press, 2014).

78 Keith Ward, *More Than Matter* (Lion, 2010).

79 Alvin Plantinga, "Materialism and Christian Belief", in P. Van Inwagen and D.W. Zimmerman (editors), *Persons: Human and Divine* (Oxford University Press, 2007), p 99-141.

80   J.P. Moreland and S.B. Rae, *Body and Soul: Human nature and the Crisis in Ethics*. (InterVarsity Press, 2000).

81   M. Egnor, "A Map of the Soul", *First Things*, 20th June 2017.

82   evolutionnews.org/2016/04/wilder_penfield/ (accessed 8th January 2019).

83   Marilynne Robinson, *Absence of Mind* (Yale University Press, 2010), p 47-50.

84   G. Wald, "Life and Mind in the Universe", *International Journal of Quantum Chemistry*, 1984,26(S11):1.

85   M.A. Jeeves and D. Tutu, *The Emergence of Personhood: A Quantum Leap?* (Eerdmans, 2015), p 241.

86   "Neuroscience and the Soul—full Interview with J.P. Moreland", youtube.com/watch?v=JxlYKqmE7o0 (accessed 8th January 2019).

87   Richard Swinburne, *The Existence of God* (Clarendon Press, 2004).

88   Sam Harris, *Free Will* (Free Press, 2012), p 4-5.

89   P. Van Inwagen, *An Essay on Free Will* (Clarendon Press, 1983).

90   P. Van Inwagen and P. Westview, *Metaphysics* (Westview Press, 2015); C. Sartorio, *Causation and Free Will*, (Oxford University Press, 2016).

91   M. McKenna and D.J. Coates, "Compatibilism" in E.N. Zalta (editor), *The Stanford Encyclopedia of Philosophy*, Winter 2016 Edition, plato.stanford.edu/entries/compatibilism/ (accessed 8th January 2019).

92   For an overview, see *The Stanford Encyclopedia of Philosophy*. Winter 2016 (see above for URL),

93   P.G.H. Clarke, *All in the Mind? Challenges of Neuroscience to Faith and Ethics* (Lion Books, 2015), p 93.

94   J. Eccles, "A Unitary Hypothesis of Mind-Brain Interaction in the Cerebral Cortex" in *Proceedings of the Royal Society of London Series B, Biological Sciences*, 1990, 240 (1299), p 433-51.

95   Richard Swinburne, *Mind, Brain, and Free Will* (Oxford University Press, 2014), p 2.

96   B. Libet, E.W. Wright and C.A. Gleason, "Readiness-Potentials Preceding Unrestricted Spontaneous vs. Pre-Planned Voluntary Acts", *Electroencephalography and Clinical Neurophysiology*, 1982, 54(3):322-35; B. Libet, C.A. Gleason, E.W. Wright and D.K. Pearl, "Time of Conscious Intention to Act in Relation to Onset of Cerebral Activity (Readiness-Potential)", *Brain*, 1983, 106(3):623-42.

97   M. Matsuhashi and M. Hallett, "The Timing of the Conscious Intention to Move" in *European Journal of Neuroscience*, 2008, 28(11):2344-51; J. Miller, P. Shepherdson and J. Trevena, "Effects of Clock Monitoring on Electroencephalographic Activity: Is Unconscious Movement Initiation an Artifact of the Clock?" *Psychological Science*, 2011, 22(1):103-9.

98   M.A. Jeeves, *Minds, Brains, Souls, and Gods: A Conversation on Faith, Psychology, and Neuroscience* (InterVarsity Press, 2013), p 56-57.

99   New Scientist, *Your Conscious Mind: Unravelling the Greatest Mystery of the Human Brain* (John Murray Learning, 2017).

100 J. Miller, P. Shepherdson and J. Trevena, "Effects of Clock Monitoring on Electroencephalographic Activity: Is Unconscious Movement Initiation an Artifact of the Clock?" *Psychological Science*, 2011,22(1):103-9.

101 W. Penfield, *The Mystery of the Mind: A Critical Study of Consciousness and the Human Brain* (Princeton University Press, 1975), p 77-78.

102 M. Shermer, "How Free Will Collides with Unconscious Impulses", *Scientific American*, 16th July 2012.

103 B. Libet, "Do We Have Free Will?" *Journal of Consciousness Studies*, 1999, 6(8):54.

104 S. Harris, *Free Will* (Free Press, 2012), p 4.

105 There are, of course, different views within Christianity on the extent of human freedom. A helpful starting book would be John Lennox, *Determined to Believe* (Lion Hudson, 2018).

106 C.S. Lewis, *Surprised by Joy* (William Collins, 2012), p 260-261

107 H. Whitehouse, "Cognitive Evolution and Religion: Cognition and Religious Evolution", in J. Bulbulia, R. Sosis, E. Harris, Russell Genet, C. Genet and K. Wyman (editors), *The Evolution of Religion: Studies, Theories, and Critiques* (Collins Foundation Press, 2008), p 19-29; J.B. Stump and A.G. Padgett, *The Blackwell Companion to Science and Christianity*, (Wiley-Blackwell, 2012).

108 C. Foster, *Wired for God? The Biology of Spiritual Experience* (Hodder, 2011).

109 D.C. Dennett, *Breaking the Spell: Religion As a Natural Phenomenon* (Penguin, 2007).

110 See for example S. Atran, *In Gods We Trust: The Evolutionary Landscape of Religion* (Oxford University Press, 2002); P. Boyer, *Religion Explained: The Evolutionary Origins of Religious Thought* (Basic Books, 2001); J. Bering, *The Belief Instinct: The Psychology of Souls, Destiny, and the Meaning of Life* (W.W. Norton & Co, 2011).

111 P. Krummenacher, C. Mohr, H. Haker and P. Brugger, "Dopamine, Paranormal Belief, and the Detection of Meaningful Stimuli in *Journal of Cognitive Neuroscience*, 2010, 22(8):1670-81.

112 Richard Dawkins, *The God Delusion*, (Bantam Press, 2006).

113 Attributed to Pirsig in the Preface to *The God Delusion*, p 28. It is perhaps a paraphrase of the following from R.M. Pirsig *Lila - An Inquiry Into Morals* (Alma Books, 2011): "An insane delusion can't be held by a group at all. A person isn't considered insane if there are a number of people who believe the same way. Insanity isn't supposed to be a communicable disease. If one other person starts to believe him, or maybe two or three, then it's a religion.".

114 M.J. Murray, "Four Arguments That the Cognitive Psychology of Religion Undermines the Justification of Religious Belief" in J.A. Bulbuli (editor), *The Evolution of Religion: Studies, Theories and Critiques* (Collins Foundation Press, 2008), p 394.

115 M.J. Murray, p 395.

116 I. Pyysiäinen and M. Hauser, "The Origins of Religion: Evolved Adaptation or By-Product?" *Trends in Cognitive Sciences*, 2010, 14(3):104-9..

117 K.J. Clark and J.T. Winslett, "The Evolutionary Psychology of Chinese Religion: Pre-Qin High Gods as Punishers and Rewarders", *Journal of the American Academy of Religion*, 2011, 79(4):928-60.

118 J. Dominic and B. Jesse, "Hand of God, Mind of Man: Punishment and Cognition

in the Evolution of Cooperation", *Evolutionary Psychology*, 2006, 4(1); D.P.P. Johnson and O. Kruger, "The Good of Wrath: Supernatural Punishment and the Evolution of Cooperation" in *Political Theology*, 2004, 5:159-76.

119 P. Boyer, *Religion Explained: The Evolutionary Origins of Religious Thought*, (Basic Books, 2001); P. Boyer, "Are Ghost Concepts 'Intuitive', 'Endemic' and 'Innate'?" in *Journal of Cognition and Culture*, 2003,3(3):233-43; P. Boyer, *The Naturalness of Religious Ideas: A Cognitive Theory of Religion* (University of California Press, 1994).

120 Alvin Plantinga, *Where the Conflict Really Lies: Science, Religion, and Naturalism* (Oxford University Press, 2011).

121 John C. Lennox, "Belief in God in 21st Century Britain", National Parliamentary Prayer Breakfast, 25th June 2013.

122 J.S. Wilkins and P.E. Griffiths, "Evolutionary Debunking Arguments in Three Domains: Fact, Value, and Religion" in G.W. Dawes and J. Maclaurin (editors), *A New Science of Religion* (Routledge, 2013), p 133-46.

123 N.G. Waller, B.A. Kojetin, T.J. Bouchard, D.T. Lykken and A. Tellegen, "Genetic and Environmental Influences on Religious Interests, Attitudes, and Values: A Study of Twins Reared Apart and Together" in *Psychological Science* 2017, 1(2):138-42; L. Eaves, "Genetic and Social Influences on Religion and Social Values" in M.A. Jeeves (editor), *From Cells to Souls—and Beyond: Changing Portraits of Human Nature* (W.B. Eerdmans, 2004), p 102-22.

124 J.L. Barrett, "Toward a Cognitive Science of Christianity" in J.B Stump and A.G. Padgett (editors), *The Blackwell Companion to Science and Christianity* (Wiley-Blackwell, 2012), p 329.

125 Barrett, p 323.

126 plato.stanford.edu/entries/altruism-biological/ (accessed 8th January 2019).

127 plato.stanford.edu/entries/altruism-biological/#ButItReaAlt (accessed 8th January 2019).

128 Barrett, p 327.

129 John Newton, "Amazing Grace", 1779.

130 W. Penfield and T.B. Rasnussen, "The Cerebral Cortex of Man" (Macmillan,1950) in G.R. Habermas and J.P. Moreland, *Beyond Death: Exploring the Evidence for Immortality* (Wipf & Stock, 2004), p 168.

131 www.issr.org.uk/issr-statements/neuroscience-religious-faith/ (accessed 8th January 2019).

132 William James, *The Varieties of Religious Experience* (Longmans, Green & Co. 1902).

133 *Oxford Dictionary of Philosophy* (OUP, 2008).

134 V.S. Ramachandran, "The Neural Basis of Religious Experiences", Society for Neuroscience Conference Abstracts 1997:1316.

135 C.R. Albright and J.B. Ashbrook, *Where God Lives in the Human Brain* (Sourcebooks, 2001).

136 C. Foster, *Wired for God? The Biology of Spiritual Experience* (Hodder, 2011), p 56.

137 V.S. Ramachandran, *The Tell-Tale Brain: Unlocking the Mystery of Human Nature*

(Windmill, 2012), p 116.

138 A. Newberg, A. Alavi, M. Baim, M. Pourdehnad, J. Santanna and E. d'Aquili, "The Measurement of Regional Cerebral Blood Flow During the Complex Cognitive Task of Meditation: A Preliminary SPECT Study" in *Psychiatry Research: Neuroimaging*, 2001, 106(2):113-22; A.B. Newberg, N.A. Wintering, D. Morgan and M.R. Waldman, "The Measurement of Regional Cerebral Blood Flow During Glossolalia: A Preliminary SPECT study", *Psychiatry Research: Neuroimaging Psychiatry Research: Neuroimaging*, 2006, 148(1):67-71.

139 W.S. Brown, "Neuroscience and Religious Faith", The International Society for Science and Religion [Internet], 2017, www.issr.org.uk/issr-statements/neuroscience-religious-faith (accessed 8th January 2019).

140 plato.stanford.edu/entries/religious-experience/#TypRelExp (accessed 8th January 2019).

141 N.T. Wright, "Mind, Spirit, Soul and Body: All for One and One for All Reflections on Paul's Anthropology in his Complex Contexts". Society of Christian Philosophers: Regional Meeting, 18th March 2011, Fordham University, New York: ntwrightpage.com/2016/07/12/mind-spirit-soul-and-body (accessed 8th January 2019).

142 A.H. Ross, *Who Moved the Stone?* (Faber & Faber, 1930); G.R. Habermas, A. Flew and T.L. Miethe, *Did Jesus Rise from the Dead?:The Resurrection Debate* (Wipf & Stock, 2003).

143 B. Holmes, "Why Be Conscious: The Improbable Origins of our Unique Mind", *New Scientist*, 10th May 2017.

144 "Material Girl" by Peter Brown and Robert Rans, performed by Madonna (Sire Records, 1985).

145 Similar to the radio example used by Justin Barrett in J.L. Barrett, "Toward a Cognitive Science of Christianity", in J.B Stump and A.G. Padgett (editors), *The Blackwell Companion to Science and Christianity* (Wiley-Blackwell, 2012), p 329.

146 D. Bennett, *A War of Loves* (Hodder, 2018).

147 Steven Pinker ,"The Brain: The Mystery of Consciousness", *Time*, 29th January 2007.

148 N.T. Wright, *Surprised by Hope* (SPCK, 2011).

ACKNOWLEDGEMENTS

Many people have played a role in getting this book ready. Thank you to the team at TGBC, and especially my editor, Tim Thornborough. I am grateful to many at the Oxford Centre for Christian Apologetics, including Amy Orr-Ewing, Nancy Gifford, John Lennox, Tom Price, Max Baker-Hytch, Ben Page, Simon Wenham and Andy Ruffhead. Beyond OCCA, I am also very grateful to Kate Blanshard.

This book would not have been written without such a generous family. Thank you, Abby and Ethan, for understanding why I was squirrelled away writing a brain book! I am deeply grateful to my husband, Conrad, for believing that this book was possible, and for his unrelenting encouragement and help in making it happen. But my biggest thanks must go to the one who makes all things possible, the Lord Jesus Christ.

the good book
COMPANY

Thanks for reading this book. We hope you enjoyed it, and found it helpful.

Most people want to find answers to the big questions of life: Who are we? Why are we here? How should we live? But for many valid reasons we are often unable to find the time or the right space to think positively and carefully about them.

Perhaps you have questions that you need an answer for. Perhaps you have met Christians who have seemed unsympathetic or incomprehensible. Or maybe you are someone who has grown up believing, but need help to make things a little clearer.

At The Good Book Company, we're passionate about producing materials that help people of all ages and stages understand the heart of the Christian message, which is found in the pages of the Bible.

Whoever you are, and wherever you are at when it comes to these big questions, we hope we can help. As a publisher we want to help you look at the good book that is the Bible because we're convinced that as we meet the person who stands at its centre—Jesus Christ—we find the clearest answers to our biggest questions.

Visit our website to discover the range of books, videos and other resources we produce, or visit our partner site www.christianityexplored.org for a clear explanation of who Jesus is and why he came.

Thanks again for reading,

Your friends at The Good Book Company

thegoodbook.com | thegoodbook.co.uk
thegoodbook.com.au | thegoodbook.co.nz
thegoodbook.co.in

WWW.CHRISTIANITYEXPLORED.ORG

Our partner site is a great place to explore the Christian faith, with powerful testimonies and answers to difficult questions.